HAZARDOUS WASTE INCINERATION CALCULATIONS

HAZARDOUS WASTE INCINERATION CALCULATIONS

Problems and Software

JOSEPH P. REYNOLDS
Department of Chemical Engineering
Manhattan College

R. RYAN DUPONT
Department of Environmental and Civil Engineering
Utah State University

LOUIS THEODORE
Department of Chemical Engineering
Manhattan College

A Wiley-Interscience Publication

JOHN WILEY & SONS, INC.

New York / Chichester / Brisbane / Toronto / Singapore

In recognition of the importance of preserving what has been written, it is a policy of John Wiley & Sons, Inc., to have books of enduring value published in the United States printed on acid-free paper, and we exert our best efforts to that end.

Library of Congress Cataloging in Publication Data:

Reynolds, Joseph P.
Hazardous waste incineration calculations : problems and software / Joseph P. Reynolds, R. Ryan Dupont, Louis Theodore.
p. cm.
"A Wiley-Interscience publication."
Includes bibliographical references.
1. Hazardous wastes—Incineration—Problems, exercises, etc.
2. Hazardous wastes—Incineration—Computer programs. I. Dupont, R. Ryan. II. Theodore, Louis. III. Title.
TD1062.R49 1991
628.4'2—dc20 91-11047
ISBN 0-471-50782-2 CIP

Printed in the United States of America

10 9 8 7 6 5 4 3 2 1

Contributors

Dr. Stuart Batterman
Dept. of Civil Engineering
Texas A&M University
College Station, Texas

Dr. Kumar Ganesan
Dept. of Environmental Engineering
Montana College of Mineral
Science & Technology
Butte, Montana

Dr. Howard E. Hesketh
Dept. of Mechanical Engineering
Southern Illinois University
Carbondale, Illinois

Dr. J. M. Hughes
Dept. of Civil Engineering
Virginia Tech
Blacksburg, Virginia

Dr. Steve E. Hrudey
Dept. of Environmental & Civil Engrg.
University of Alberta
Edmonton, Alberta, Canada

Dr. Sonia Kreidenweis
Dept. of Chemical Engineering
San Jose State University
San Jose, California

Dr. Soon-Sik Lim
Dept. of Chemical Engineering
Youngstown State University
Youngstown, Ohio

Dr. James H. McMicking
Dept. of Chemical Engineering
Wayne State University
Detroit, Michigan

Dr. W. Lamar Miller
Dept. of Environmental Engrg. Sciences
University of Florida
Gainesville, Florida

Dr. Mark J. Rood
Dept. of Civil Engineering
University of Illinois
Urbana, Illinois

Dr. Dennis Ryan
Dept. of Chemistry
Hofstra University
Hempstead, New York

Dr. Bruce M. Thomson
Dept. of Civil Engineering
University of New Mexico
Albuquerque, New Mexico

Dr. Cristos Tsiligiannis
City College of the City
University of New York
Chemical Engineering Department
New York, New York

Dr. Frank L. Worley Jr.
Dept. of Chemical Engineering
University of Houston
Houston, Texas

Dr. Jy S. Wu
Dept. of Civil Engineering
University of North Carolina
at Charlotte
Charlotte, North Carolina

Dr. Ronald F. Wukasch
Dept. of Civil (Environmental) Engineering
Purdue University
West Lafayette, Indiana

Contents

Preface **ix**

PART ONE PROBLEM STATEMENTS

1. **Basic Concepts** **3**
2. **Regulatory Concerns** **5**
3. **Stoichiometry** **9**
4. **Thermochemistry** **12**
5. **Incinerators** **18**
6. **Waste Heat Boilers/Quenchers** **22**
7. **Air Pollution Control Equipment** **24**
8. **Risk Analysis** **31**
9. **Monitoring** **36**

PART TWO PROBLEM SOLUTIONS

1. **Basic Concepts** **41**
2. **Regulatory Concerns** **46**
3. **Stoichiometry** **57**
4. **Thermochemistry** **67**
5. **Incinerators** **93**
6. **Waste Heat Boilers/Quenchers** **108**
7. **Air Pollution Control Equipment** **115**
8. **Risk Analysis** **131**
9. **Monitoring** **140**

PART THREE *USER'S GUIDE* FOR THE *HWI* SOFTWARE PACKAGE

1. **Introduction to the *HWI* Software Package** **145**
2. **Preparing to Use *HWI*** **147**
3. **The Program *HWITRL*** **152**
4. **The Program *HWI*** **156**
5. **The Program *HWISET*** **176**

PART FOUR *REFERENCE MANUAL* FOR THE *HWI* SOFTWARE PACKAGE

1. **Introduction** **189**
2. **Calculation Concepts** **190**
3. **Program Details** **224**
4. **Chlorobenzene/Sulfur Mixture Example** **230**
5. **Chlorobenzene/DDT/Water Mixture Example** **240**

References **249**

Preface

The engineering profession has recently expanded its societal responsibilities to include the management of hazardous wastes, with particular emphasis on control by incineration. Increasing numbers of engineering personnel are being confronted with problems in this crucial area. Since the problem of hazardous wastes is a relatively new concern, environmental engineers of today and tomorrow must develop a proficiency and an improved understanding of the incineration of hazardous wastes in order to cope with these challenges.

This book, plus the software contained on the computer diskette, represent an attempt to help meet these challenges. The book consists of four main parts. **Part 1** contains almost 100 problems dealing with hazardous waste incineration and related areas (e.g., quenchers, air pollution control equipment, risk analysis, etc.) and **Part 2** contains detailed solutions to these problems. **Part 3** is a *User's Guide* which provides instructions and documentation for the *HWI* software contained on the enclosed diskette. **Part 4** is a *Reference Manua-* intended to familiarize the user with the basic principles behind the calculations performed by the *HWI* program.

Parts 1 and 2 are the result of a year-long effort on the part of a group of university professors who first met during the summer of 1988 for a College Faculty Workshop. The workshop was funded by the National Science Foundation's Undergraduate Faculty Enhancement Program and was conducted at Manhattan College. The NSF objective of such a workshop is to "involve college faculty members in preparing course materials and . . . in testing the effectiveness of these science curricular innovations for implementation." During the award-selection process, the Foundation gives priority to "the development of more efficient and effective educational procedures in newly emerging, interdisciplinary, and problem-relevant subjects." The main objective of the Manhattan College Workshop was the development of a problem workbook in the hazardous waste incineration field. During the June 1988 meeting, which was of two weeks duration, each faculty member was required to generate six meaningful applications-oriented problems. The 1988–1989 academic school year afforded most of the faculty an opportunity to classroom test their own problems, at a minimum, plus many of the other problems. A three-day follow-up session held during June 1989 at Manhattan College was utilized to revise, update, and edit the workbook and finalize production and distribution plans.

The HWI software package, described and documented in Parts 3 and 4, resulted from a project sponsored by the U.S. Environmental Protection Agency's Air Pollution Training Institute. (The project officer was Charles Pratt.) The objectives of this project were to (1) develop rigorous calculational procedures for the design and analysis of incinerators, and (2) to incorporate these procedures into an easy-to-use computer program. These procedures include:

1. Thermochemical calculations, which relate operating incinerator temperature, excess air, and feed (including not only waste but also possible fuel requirements) heating value;
2. Stoichiometric calculations, which yield the composition and flow rates of gaseous emissions from incinerators burning (hazardous) waste; and
3. Incinerator design.

The overall purpose of the *HWI* Software Package is to provide a convenient method of estimating the effects associated with (1), (2) and (3) above. For category (1), equations have been developed that describe the interrelationships among such variables as: inlet temperature, outlet temperature, heat (radiation) loss, excess air, type of fuel, quantity of fuel, and waste physical and chemical properties. These results have been extended to include pollutant emission calculations (2) and incinerator physical design (3).

Besides the *HWI* program, which performs the actual calculations, the diskette contains two other programs. The *HWITRL* program is a tutorial intended to instruct the first-time user in the operation of the *HWI* program. *HWISET* is a utility program for use with HWI. All three programs are described in detail in Part 3.

The authors wish to acknowledge both the National Science Foundation and the U.S. Environmental Protection Agency, without whose support the project would not have come to fruition. Our sincere gratitude is also due to Mrs. Ann Kaptanis for her dedicated and patient efforts in typing the manuscript.

JOSEPH P. REYNOLDS
R. RYAN DUPONT
LOUIS THEODORE

Riverdale, New York
Logan, Utah
April 1991

HAZARDOUS WASTE INCINERATION CALCULATIONS

PART 1

PROBLEM STATEMENTS

Part 1 contains almost 100 problems related either directly or indirectly to the hazardous waste incineration field. These problems are organized into nine categories, each category comprising one chapter in this part. The categories are: Basic Concepts (BC), Regulatory Concerns (R), Stoichiometry (S), Thermochemistry (T), Incinerators (I), Waste Heat Boilers/Quenchers (WHB/Q), Air Pollution Control Equipment (APCE), Risk Analysis (RA), and Monitoring (M). The solution to each problem can be found in Part 2 in a corresponding chapter with the same number and title as the problem statement. Each solution is also identified with the same letters and number used for the corresponding problem statement in Part 1. For example, the solution to Problem RA-5 in Chapter 8 is identified as Solution RA-5 and can be found in Chapter 8 of Part 2.

CHAPTER ONE

Basic Concepts

BC-1. Avogadro's number is 6.02×10^{23} particles/gmol. The molar volume of an ideal gas at 0 °C and 1 atm of pressure is 22.4 L. (dr)

a. How many molecules are there in 1.00 lbmol of SO_2?
b. What is the volume in ft^3 of one lbmol of an ideal gas at 60 °F and 1 atm of pressure?
c. What is the value of the ideal gas constant in units of psi-ft^3/lbmol-°R?
d. The flame temperature in a hazardous waste incinerator is 2460 °R. What is this temperature in kelvins? (Hint: °C = (°F-32) x 9/5)

BC-2. A flue gas (MW=30, dry basis) is being discharged from a scrubber at 180°F (dry bulb) and 125 °F (wet bulb). The gas flow rate on a dry basis is 10,000 lb/h. (jhm)

a. What is the dew point of the wet gas?
b. What is the mass flow rate of the wet gas?
c. What is the actual volumetric flow rate of the wet gas?

BC-3. Based on the data given below for a chemical holding tank, answer the following questions: (rfw)

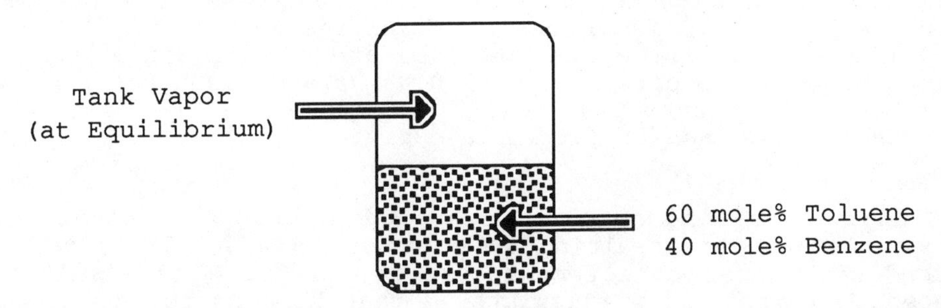

Tank pressure = 1 atm
Tank temperature = 60 °F
Toluene vapor pressure at 60 °F = 16 mm Hg
Benzene vapor pressure at 60 °F = 60 mm Hg

a. What is the concentration of hydrocarbon in the vapor phase?
b. Is the vapor phase flammable or explosive?
c. What is the NHV of the vapor in Btu/ft^3?
d. Is the mixture "lean" or "rich", i.e., does it require more fuel or more air for combustion if it is to be incinerated?

BC-4. The heat of formation of ethylene glycol is -109 kcal/gmol. Estimate the GROSS and NET heat of combustion of ethylene glycol in Btu/lb. Use heat of formation data in Appendix H. (rfw).

BC-5. Process considerations require pH control in a 50,000 gal storage tank used for incoming waste at a hazardous waste incinerator. Normally, the tank is kept at neutral pH; however, operation can tolerate pH variations from 6 to 8. Waste arrives in 5000 gal shipments. Assume that the tank is completely mixed, contains 45,000 gal when the shipment arrives, the incoming acidic waste is fully dissociated, and that there is negligible buffering capacity in the tank. (sb).

a. What is the pH of the most acidic waste shipment that can be handled without neutralization?
b. What is the pH of the most acidic waste shipment that can be handled without neutralization if the storage tank volume is 100,000 gal, and contained 95,000 gal of neutral waste when the shipment arrived?

BC-6. An incinerator design which has been successfully used in Houston, TX (zero ft above mean sea level), is to be used in Albuquerque, NM (5,200 ft above mean sea level). To insure proper performance of the incinerator, it is necessary to correct for the change in altitude since the density of the combustion air will decrease. Assume P = 14.7 psi, T = 60 °F, ρ = 0.0756 lb/ft^3 in Houston. (bmt)

a. Calculate the theoretical air requirements in cfm to burn 10^5 lb/d of chlorobenzene (C_6H_5Cl) at sea level.
b. Calculate the required air flow rate (cfm) to achieve complete combustion in Albuquerque. Assume the pressure and temperature in Albuquerque are 12.1 psi and 501.5 °R, respectively.
c. Calculate the allowable particulate concentration (gr/acfm) in Albuquerque which corresponds to 0.08 gr/acfm in Houston. The allowable particulate concentration in Albuquerque should be calculated based on the same total mass of particulates.

BC-7. Sludge containing mercury is burned in an incinerator (mercury feed rate 9.2 lb/h). The resulting 500°F product (40,000 lb/h, gas, MW = 32) is quenched with water to a temperature of 150 °F. The resulting stream is filtered to remove all particulates. What happens to the mercury? Assume the process pressure is 14.7 psi and that the vapor pressure of Hg at 150°F is 0.005 psi. (fw)

BC-8. Tests indicate that waste arriving in 50 gal drums to a hazardous waste incineration facility have a mean Pb content of 20 ppm with a standard deviation of 15 ppm. Drums are unloaded into a 350 gal receiving tank. The facility is required to keep the Pb concentration entering the incinerator at or below 25 ppm. Assume that the Pb contents of the drums are not correlated and that the tank is nearly full. (sb)

a. What is the probability that the Pb content in any truck exceeds 50 ppm?
b. What is the probability that the Pb content in the receiving tank exceeds 25 ppm?
c. What size should the receiving tank be to ensure, with 95% confidence, that the incinerator burns a waste with a mean Pb concentration below 25 ppm?
d. What are the limitations of this analysis, and how could these limitations be accommodated?

CHAPTER TWO

Regulatory Concerns

R-1. Briefly describe the major strong points of the permit process for incinerator systems under RCRA. Include the perspective of the public, the regulatory agency, and the permittee. (lm)

R-2. The term liability is closely tied to the system of values and ethics which we have developed in this country. In concise language, define and give an example of the terms **liability, strict liability,** and **joint and several liability**. Explain how our interpretation of liability affects hazardous waste incineration permitting, application and design. (lm)

R-3. In passing the Hazardous and Solid Waste Act amendments (HSWA) of 1984, the Congress required hazardous waste generators to certify that they had a program in place to reduce the volume and/or toxicity of the waste they generated or managed. Explain why the Congress did not include a corresponding enforcement section in the law. (lm)

R-4. A major problem facing engineers working for or with industry today is the necessity of being able to design and implement systems which will serve their intended purposes throughout their intended design life. This is made extremely difficult, especially within the environmental engineering arena, when the engineer is faced with constantly changing regulatory requirements. Explain how the Congress and EPA attempt to minimize the effect of this problem, while at the same time protect the environment from adverse effects of hazardous waste incineration. (lm)

R-5. A hazardous waste incinerator is burning an aqueous slurry of soot (carbon) with the production of a small amount of fly ash. The waste is 70% water by mass and is burned with 0% excess air (EA).

The flue gas generated contains 0.30 gr of particulates in each 8.0 ft^3 (actual) at 580 °F. Calculate the particulate concentration in the flue gas in gr/acf, in gr/scf, in gr/dscf and in gr/dscf corrected to 50% EA. If the regulations require that particulate emissions be less than 0.08 gr/dscf corrected to 50% EA, is this incinerator in compliance or must additional particulate control measures be taken? (dr)

Assume that when the flue gas passes through a waste heat boiler no water condensation occurs.

R-6. A compliance stack test of an incinerator yields the results below. Determine whether the incinerator meets the state particulate standard of 0.05 gr/dscf. Estimate the amount of particulate matter escaping the stack, and indicate the molecular weight of the stack gas. Use standard conditions of 70 °F and 1 atm pressure. (kg)

Volume Sampled	=	35 dscf
Diameter of Stack	=	2 ft
Pressure of Stack Gas	=	29.6 in Hg
Stack Gas Temperature	=	140 °F
Mass of Particulate Collected	=	0.16 g
% Moisture in Stack Gas	=	7% (by volume)
% O_2 in Stack Gas (dry)	=	7% (by volume)
% CO_2 in Stack Gas (dry)	=	14% (by volume)
% N_2 in Stack Gas (dry)	=	79% (by volume)
Pitot Tube Factor (k)	=	0.85

R-6 (Continued).

Pitot tube measurements made at eight points across the diameter of the stack provided values of: 0.3, 0.35, 0.4, 0.5, 0.5, 0.4, 0.3, and 0.3 inches of H_2O.

Use the following equations for S-type pitot tube velocity, v (m/s), measurements:

$$v = k\sqrt{2\ g\ H}$$

$$= k\sqrt{2\ g\ \frac{\rho_1}{\rho_a}\ (0.0254)\ h}$$

where
- g = gravitational acceleration 9.81 m/s^2
- H = fluid velocity head (inches of H_2O)
- ρ_1 = density of manometer fluid (1000 kg/m^3)
- ρ_a = density of flue gas (1.084 kg/m^3)
- h = mean pitot tube reading (inches of H_2O)

R-7. A coal process unit waste, containing carbon and free water, is being incinerated by ABC Waste Disposal, Inc. The flue gas leaving the incinerator is at a temperature of 450 °F following a waste heat recovery unit, and contains a particulate loading of 0.06 gr/acf. The flue gas analysis shows the following composition:

% O_2 in Stack Gas = 12.5%
% CO_2 in Stack Gas = 12.5%
% N_2 in Stack Gas = 50.0%
% H_2O in Stack Gas = 25.0%

Determine the outlet particulate loading based on a dry, standard atmosphere, corrected to 50% EA. Is the process in compliance with current particulate regulations? Use standard conditions of 1 atm pressure and a temperature of 68 °F. (flw)

R-8. Your office has received a request for a permit to incinerate a mixture of p-dichlorobenzene (40% by weight), coal dust (40% by weight, particle diameter ≤ 5μm), and contaminated #2 fuel oil (20% by weight). The incinerator is followed by a quencher and a packed bed scrubber. Experience has shown that the quencher will remove 90% of the HCl in the gas stream. The system is operated at 50% EA. The following elemental composition data are also available to you for the waste/fuel feed to the incinerator: (jmh)

Element	Coal	Oil
C	79.3	68.8
H	3.8	7.8
Cl		10.0
O	5.3	2.0
S	1.1	1.5
N	1.0	1.7
Ash	9.5	8.2

The specific gravity of the feed streams is: 1.53 for the p-DCB, 2.3 for the coal, and 0.75 for the #2 fuel oil.

R-8 (Continued).

The equilibrium constant, K_p, for the HCl $\Leftrightarrow Cl_2$ equilibrium reaction is given as:

$$2\ HCl + 0.5\ O_2 \leftrightarrow H_2O + Cl_2$$

$$K_p = A\ e^{B/T}$$

where
$A = 0.23 \times 10^{-3}/(atm)^{0.5}$
$B = 7340$ K
T = temperature (K)

The incinerator to be used is 5 ft in diameter and 25 ft long. The heat release rate should not exceed 25,000 Btu/h-ft^3. Heat loss from the incinerator is estimated to be 8%. The quencher spray rate is to be set so that the quencher exit temperature is 87 °C.

a. Calculate the NHV of the feed mixture in Btu/lb using Dulong's equation.
b. Calculate the maximum feed rate, in gpm, which can be incinerated.
c. Estimate the incinerator exhaust temperature in °F using the following empirical expression:

$$T = 60 + \frac{NHV}{(0.3)\ [1 + (1 + EA)\ (7.5 \times 10^{-4})\ (NHV)]}$$

d. What would be the Cl_2 concentration in ppm leaving the combustion chamber?
e. What is the required packed bed scrubber HCl removal efficiency to meet the 4 lb/h emission standard?
f. What is the residence time in the incinerator in s?
g. Briefly critique this proposal from the standpoint of technology, emissions, energy consumption and economy.

R-9. A mixture of trichloroethylene, tetrachloroethylene, dichlorofluoromethane, and phthaloyl chloride in #2 fuel oil (1% sulfur) is fired during a trial burn in a liquid injection incinerator. The facility is equipped with a quench tower, venturi scrubber and packed bed caustic scrubber.

Each of the four organic compounds makes up 5% (by weight) of the waste/fuel feed of 5000 lb/h. The excess air is 35%. The stack gas flow rate, corrected to 7% oxygen and standard conditions (1 atm, 25°C), is 21,300 dscfm. (jmh)

a. Calculate the destruction and removal efficiency (DRE) for each hazardous constituent using the flue gas measurements below.

C_2HCl_3	0.03 ppm
C_2Cl_4	0.02 ppm
$CHCl_2F$	0.30 ppm
$C_3H_4Cl_2O$	1.8 ppm

b. Use DuLong's equation to calculate the NHV of the mixture in Btu/lb. The NHV of # 2 fuel oil is 18,650 Btu/lb.

R-10. A state incinerator emission limit requires 99% HCl control and allows 0.07 gr/dscf at 68°F, corrected to 50% EA. An incinerator is to burn 5 ton/h hazardous waste containing 2% Cl, 80% C, 5% inerts, and the balance H_2O (by weight). (heh)

a. Calculate the maximum mass emission rate of equivalent HCl in lb/h that may be emitted.
b. Calculate the maximum mass emission rate of particulates in lb/h that may be emitted. What is the actual particulate mass emission rate if all the inerts are emitted from the stack as fly ash.
c. Determine the combustion efficiency of this incinerator if a stack test indicates that the flue gas contains 12% CO_2 and 20 ppm CO.

R-11. A hazardous waste incinerator is burning a waste mixture with 50% excess air at 2100°F with a residence time of 2.5 s. The stack gas flow rate was determined to be 14,280 dscfm. The composition of contaminants in the stack gas is given below: (jsw)

Compound	Inlet (lb/hr)	Outlet (lb/hr)
toluene	860	0.20
chlorobenzene	450	0.02
dichlorobenzene	300	0.03
HCl		4.2
particulates		9.65
Pb		0.05 lb/100 lb (TOC)

a. Calculate the DREs for toluene, chlorobenzene, dichlorobenzene and HCl.
b. Calculate the discharge concentrations of Pb and particulates.
c. Is the unit in compliance with present federal regulations? Assume chlorobenzene and dichlorobenzene are the POHCs for this incinerator.

CHAPTER THREE

Stoichiometry

S-1. The general formula for the alkyl-dichlorobenzenes is:

$$C_nH_{2n-8}Cl_2, \text{ where } n > 6$$

Write a balanced, general chemical equation for the combustion of alkyldichlorobenzenes. (dr)

S-2. A waste mixture to be incinerated in a liquid injection incinerator has the following composition per lb of waste:

0.109 lb H	0.732 lb C
0.102 lb O	0.035 lb Cl
0.022 lb water	

The incineration system is configured as shown below:

Waste Mixture + Dry Air ⇒ Incinerator

⇓

Na_2CO_3 ⇒ Wet Scrubber ⇒ Flue Gas

a. If dry air is used for combustion, how many lb of water will be present in the flue gas, before the scrubber, per lb of waste?
b. For the same waste mixture, calculate the stoichiometric volume of oxygen (at 60 °F and 1 atm) required per lb of waste.
c. Soda ash (Na_2CO_3) is used in a wet scrubber to neutralize the HCl produced during combustion. How many lbmol of soda ash are needed per pound of waste for exact neutralization? (dr)

S-3. A hazardous waste incinerator has been burning a certain mass of dichlorobenzene ($C_6H_4Cl_2$) per hour and the HCl produced was neutralized with soda ash (Na_2CO_3). If the incinerator switches to burning an equal mass of mixed tetrachlorobiphenyls ($C_{12}H_6Cl_4$), by what factor will the consumption of soda ash be increased? (dr)

S-4. The products of the reaction of chlorine with an equal number of moles of water are passed through a scrubber for complete HCl removal, i.e.,

$$Cl_2 + H_2O \rightarrow 2\ HCl + 0.5\ O_2$$

Ninety-nine percent of the scrubber outlet gas is recycled as indicated in the diagram below. The recycle gas stream is mixed with the fresh feed, and there is a negligible loss of HCl in the release gas. (ct)

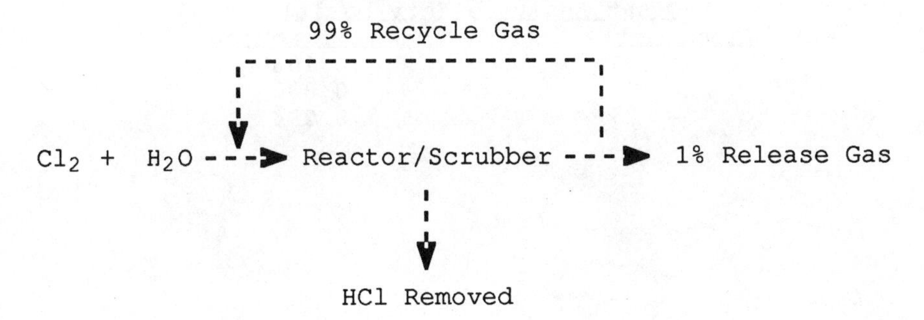

S-4 (Continued).

a. For what recycle ratio is the Cl_2 mole fraction of the release gas 0.1?
b. What are the mole fractions of the reactor products?
c. What is the reactor temperature if the pressure is 1 atm?

Note: The change in equilibrium constant for the reaction shown above, as a function of temperature, can be approximated by:

$$-\ln(K_p) = \frac{7048.7}{T} + 0.151 \ln(T) - 9.06 \times 10^{-5} T - 27{,}141 T^{-2} - 8.09$$

where T is in K.

S-5. An incinerator rated at 75 x 10^6 Btu/h heat input is to destroy 3 t/h of sludge initially at 60°F. The NHV of the sludge is 1000 Btu/lb. The primary combustion chamber operates at 2000 °F and there is no secondary combustion unit. Estimate the amount of natural gas (assume CH_4) of 1000 Btu/ft^3 heating value that is required, and the composition of the exhaust gas. The sludge contains 40% free water, 30% C, 6% H, 2% Cl, 8% O, 4% N, and 10% inerts (by mass). Neglect heat losses, assume complete combustion, and assume stoichiometric air is supplied. Use a specific heat (Cp) of 1.5 Btu/lb for the inert fraction of the waste. (heh)

S-6. A company is planning to combust 1000 lbmol/h of 90 mol% C_6H_5Cl and 10 mole% CH_2Cl_2 at 100% EA. There is 5% by weight of suspended inert material in the waste stream. Ambient temperature, relative humidity, and pressure are 70 °F, 65%, and 1 atm, respectively. Standard conditions are 68°F and 1 atm. Pertinent data can be found in Appendix A. (mjr)

a. What is the mole fraction of HCl in the flue gas if all the Cl reacts to form HCl?
b. What is the minimum collection efficiency of a particulate control device to meet a particulate emission standard of 0.08 gr/dscf corrected to 50% EA?

S-7. A waste mixture containing 55 weight% chlorobenzene (C_6H_5Cl, MW = 112.5) and 45 weight% DDT ($C_{14}H_9Cl_5$, MW = 177.5) is burned at a rate of 6000 lb/h at a temperature of 2500 °F. An Orsat Analysis of the stack is given below. The flow rate for the dry gas is 132,346 acfm. For the conditions stated, calculate the amount of excess air that is present during the combustion process, the acfm of the wet flue gas (including HCl), the scfm of the wet flue gas, and the flue gas mass flow rate. It may be assumed that the gas flow rate has been corrected for humidity. (jhm)

Orsat Analysis (dry basis)

Component	mole fraction
CO_2	0.106
N_2	0.800
O_2	0.094

S-8. For the packed bed absorber shown, calculate the mass flow rate of NaOH required in gal/h. (smk)

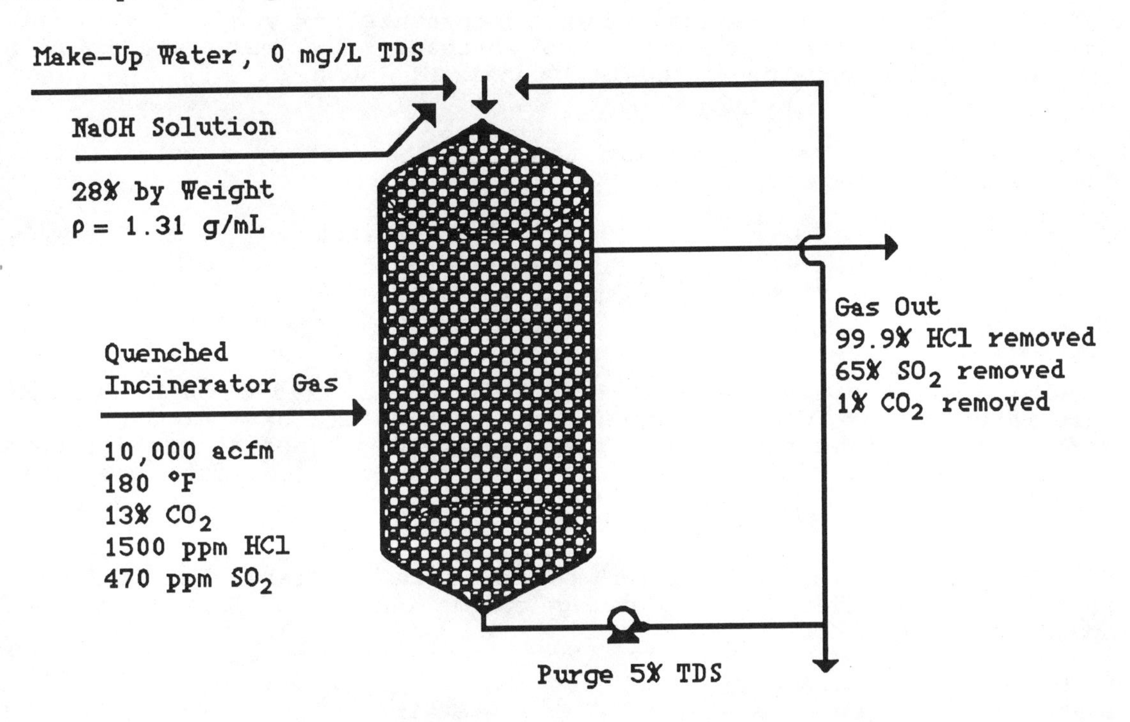

S-9. Combustion of a hazardous waste produces 10,000 acfm of flue gas at 750 °F and 1 atm. HCl and SO_2 concentrations are 20,000 ppm and 350 ppm, respectively. HCl must be controlled to 99% collection efficiency or 4 lb/h. SO_2 emissions are controlled at 70% removal efficiency. A spray dryer is used to control the HCl and SO_2 emissions. $Ca(OH)_2$ is the sorbent that will react with HCl and SO_2 to form $CaCl_2$ and $CaSO_4$, respectively. Assume that it is necessary to provide 10% and 30% excess lime feed for the required HCl and SO_2 removal, respectively. (mjr)

a. What is the required feed rate of $Ca(OH)_2$?
b. What is the total mass production rate of solids from the spray dryer? Assume that the excess lime solids in the spray dryer are $Ca(OH)_2$.

S-10. An incinerator burns mercury contaminated waste. The waste material has an ash content of 1%. The solid waste feed rate is 1000 lb/h and the gas flow rate is 20,000 dscfm. It is reported that the average mercury content in the particulates was 2.42 µg/g when the vapor concentration was 0.3 mg/dscm. For the case where incinerator emissions meet the particulate standard of 0.08 gr/dscf (0.1832 g/dscm) with a 99.5% efficient ESP, calculate: (kg)

a. The amount of mercury bound to the fly ash which is captured in the ESP in g/d.
b. The amount of mercury leaving the stack as a vapor and with the fly ash in g/d.

CHAPTER FOUR

Thermochemistry

T-1. Calculate the theoretical flame temperature for a mixture containing 75% by weight chloroethylene and 25% propane. Estimate the molar heat capacities using data presented in Appendix E. (jhm)

T-2. A contaminated waste mixture to be incinerated by a rotary kiln has the following composition.

C	0.35 lb/lb waste
H	0.07 lb/lb waste
Water	0.45 lb/lb waste
O	0.11 lb/lb waste
S	0.02 lb/lb waste

The waste feed rate is 5,000 lb/h. Natural gas (100% CH_4) is used as auxiliary fuel for the combustion process at a rate of 1 lb CH_4/lb waste. What is the maximum temperature that can be maintained in the incinerator if 100% excess air is used? Neglect heat losses. Use data presented in Appendixes E and H. (ssl)

T-3. Calculate the combustion temperature (adiabatic) for carbon tetrachloride under the following conditions (Use data presented in Appendix F): (sw)

a. Pure oxygen combustion.
b. Theoretical flame temperature (0% excess air).
c. 100% excess air.

T-4. Calculate the theoretical flame temperature of a hazardous waste mixture containing 30% cellulose, 30% motor oil, 20% water and 20% inerts. Assume 5% radiant heat losses. Use 14,000 Btu/lb and 25,000 Btu/lb as the NHV of the cellulose and oil respectively. The incinerator emission gas contains 11.8% CO_2, 13 ppm CO and 10.4% O_2 (dry basis). Suggestion: Use the equation of Theodore-Reynolds shown below to predict flame temperature. (heh)

$$T = 60 + \frac{NHV}{(0.325)\ [1 + (1 + EA)(7.5 \times 10^{-4})(NHV)]}$$

where T = temperature (°F)
NHV = net heating value of the mixture (Btu/lb)
EA = excess air on a fractional basis.

Estimate EA from the following equation, where Y = dry mole% of O_2 in the combustion gas:

$$EA = \frac{0.95\ Y}{(21 - Y)}$$

T-5. A waste mixture containing CCl_4 (20% by weight) and CH_4 is to be incinerated. Calculate the following using C_p data from Appendix F: (sw)

a. The stoichiometric oxygen requirement.
b. The adiabatic combustion temperature at 0% excess air.
c. The percent of excess air to maintain an operating temperature of 2100°F.

T-6. Compare the heat of combustion of one mole of benzene (C_6H_6) with the combined heats of combustion of six moles of carbon and three moles of H_2. (ssl)

T-7. Estimate the combustion temperature of a liquid incinerator burning contaminated alcohol [50% ethyl alcohol + 50% water by weight] using 40% EA. To simplify the calculations, assume an average flue gas heat capacity for N_2, CO_2, and O_2 of 0.26 and for $H_2O(g)$ of 0.5 Btu/lb-°F. (rfw)

T-8. The incineration of chlorinated hydrocarbons can be accomplished by using supplemental energy derived from hydrocarbon waste or other high energy fuel. (flw)

a. If the combustion temperature must be 2400 °F in order to destroy carbon tetrachloride (the chlorinated waste), how much supplemental fuel as CH_4 is needed at 30% EA? Assume all C in the waste forms CO_2, while Cl is converted to HCl if adequate H is available or Cl_2 in the event that H is not available.
b. Calculate the composition (weight%) of the gas leaving the unit.

T-9. A hazardous waste incinerator burning significant quantities of bromine-containing waste produces unacceptable amounts of Br_2 in the stack gas. The bromine is formed in the kiln according to the equation:

$$2\ HBr + 1/2\ O_2 \leftrightarrow Br_2 + H_2O$$

Because the reaction is exothermic in the forward direction, a suggestion is made to try burning the waste in a plasma arc furnace at significantly higher temperatures than conventional incineration. If the equilibrium shifts to the left, then the bromine will exist as HBr which is easy to remove in an alkaline scrubber. Average heat capacities are available for some of these compounds in Appendix C. In addition, average heat capacities over the interval 0 °F to 4000 °F have been estimated for Br_2 and HBr as follows:

Mean C_p HBr = 8.289 Btu/lbmol-°R, Mean C_p Br_2 = 8.555 Btu/lbmol-°R

Use these average heat capacity values to estimate the value of the equilibrium constant for the above reaction at 4000 °F. For constant (average) heat capacities, the following form of the integrated van't Hoff equation applies:

$$\ln K = -\frac{\Delta H^\circ}{RT} + \frac{\Delta \bar{C}_p T}{RT} + \frac{\Delta \bar{C}_p \ln T}{R} + I$$

where I is an integration constant and ΔH° is standard enthalpy of reaction.

$$\Delta \bar{C}_p = \bar{C}_p(\text{products}) - \bar{C}_p(\text{reactants})$$

$$\bar{C}_p = \text{mean heat capacity over given temperature range.}$$

Based on the above estimate, what can be concluded as to whether or not higher kiln temperatures will lead to decreased Br_2 emissions? The heat of formation of HBr at 298 K is -8660 cal/gmol. The free energy of formation of HBr at 298 K is -12,720 cal/gmol. (dr)

T-10. Two wastes are received and stored in separate tanks at a liquid injection incinerator facility. The first, a sludge with an NHV of 6,000 Btu/lb, contains 2% Cd by weight. The second, a plating waste with an NHV of 8,000 Btu/lb, contains 8% Cd by weight. A minimum of 1,000 lb/h of each waste is to be incinerated. Because of pump limitations, no more than 5,000 lb/hr of each waste can be utilized.

To achieve the required DRE, the facility requires a minimum of 8,000 Btu/lb in the waste stream, which may be obtained by adding fuel oil with an NHV of 15,000 Btu/lb. The incinerator can operate with a heat rate between 25 and 40 million Btu/h. A graphical approach involving the plotting of the two principal variables, Q_{sludge} and $Q_{plating\ waste}$, may aid in the following analysis. (sb)

a. Determine the maximum amount of plating waste that can be incinerated.
b. Determine the maximum amount of sludge that can be incinerated, along with the required supplemental fuel oil necessary for this waste.
c. If no more than 150 lb/hr of Cd can be incinerated, what is the maximum amount of plating waste that can be incinerated?
d. Discuss the impact that the Cd and other constraints impose on the operation of this incineration facility.

T-11. Calculate the gross heating value (HHV) and the net heating value (NHV) of methane, chloroform, benzene(g), chlorobenzene and hydrogen sulfide, assuming water as a product of the combustion process for all compounds of concern. Compare these values with those calculated using Dulong's equation. Calculate the relative percent difference between the "true" NHVs as determined by thermodynamic calculations, and the "estimated" values calculated using Dulong's equation. Use data available in Appendix H. (bmt)

T-12. Calculate the enthalpy of the combustion of substituted chloromethanes (ΔH°_{298}), and plot this value as a function of the number of chlorine atoms in the molecule. Include methane in the comparison. Use the following compounds in your calculations:

$$CH_4(g),\ CH_3Cl(l),\ CH_2Cl_2(l),\ CHCl_3(l),\ \text{and}\ CCl_4(l)$$

Do not include the heats of vaporization of the liquid in your calculations, and assume the only end products are CO_2, $H_2O(g)$, and HCl, and for chloroform and carbon tetrachloride Cl_2. Use data available in Appendix H. (bmt)

T-13. A diesel oil spill has resulted in contamination of soil to an oil content of 2.5% by weight. Incineration of this contaminated soil is proposed using a 10,000 lb/h incinerator, with 20 lb/h fuel oil used as auxiliary fuel. The heat capacity of the soil is given as:

$$Cp_{soil} = 0.18 + 0.00006\ (T)$$

where Cp_{soil} = heat capacity of soil (Btu/lb-°F)
T = temperature (°F)

Assume using T = 2200 °F that Cp_{soil} = a constant, Cp_{water} = 0.52 Btu/lb-°F, Cp_{CO_2} = 0.27 Btu/lb-°F, Cp_{N_2} = 0.27 Btu/lb-°F, NHV_{diesel} = 18,600 Btu/lb, and $NHV_{fuel\ oil}$ = 19,500 Btu/lb.

T-13 (Continued).

Neglecting heat losses in the incinerator, but including the heat capacity of the soil, calculate the composition of the flue gas and the operating temperature of the incinerator. Use EPA tables for stoichiometric calculations. Data for the compositional mass fraction and NHV values for the diesel and fuel oils are provided below: (bmt)

	Diesel	#2 Fuel Oil
C	0.872	0.866
H	0.123	0.102
S	0.005	0.030
NHV (Btu/lb)	18,600	19,500

T-14. Write an expression for the reaction between N_2 and O_2 to form NO. Derive an equation which relates $\log(P_{NO}/P_{N_2})$ to logK, where K is the equilibrium constant for this reaction. Use the van't Hoff equation to develop an approximate expression for the dependence of $\log(P_{NO}/P_{N_2})$ on temperature. Compare this dependence with a more exact formulation based on heat capacity using the following general equation:

$$C_p = a + b\,(T) + c\,(T^{-2})$$

Plot $\log(P_{NO}/P_{N_2})$ versus T (K) from 298 K to 3000 K. Values for a, b and c may be obtained from data in Appendix F. (bmt)

T-15. Given the data in Problem T-14, predict the products formed when organic nitrogen compounds are combusted. Does the nitrogen from triethylamine go to N_2 or NO? Indicate the final nitrogen form when combusting nitrobenzene. (lm)

T-16. Estimate the temperature required to be maintained in an isothermal, plug flow incinerator to obtain 99.5% destruction of toluene and benzene when these wastes have a 0.5 s residence time in the combustion chamber. (kg)

There is no accurate procedure to estimate this required temperature. One approach is to use the relationship of Cooper and Alley (1985) which relates efficiency of destruction to the residence time through a first order rate constant as follows:

$$\eta = 1 - e^{(-kt)} \qquad (1)$$

where
η = fractional efficiency
k = first order rate constant for compound combustion (1/s)
t = residence time of gas in incinerator (s)

The first order rate constant, k, can be described by an Arrhenius expression of the form:

$$k = A\,e^{(-E/RT)} \qquad (2)$$

where
E = activation energy (cal/gmol)
R = ideal gas constant = 1.987 cal/gmol-K
T = absolute temperature (K)

T-16 (Continued).

The pre-exponential factor, A, can be estimated from the following equation:

$$A = \frac{P\ m_{O_2}\ C\ S}{R'} \qquad (3)$$

where
P = absolute pressure (atm)
m_{O_2} = mole fraction of O2 in incinerator
C = molecular collision rate factor
R' = ideal gas constant = 0.08205 L-atm/gmol-K
S = stearic factor = 16/molecular weight of gas

The activation energy, E (kcal/mol), has been correlated with the molecular weight of the compound by the relationship shown in Equation 4:

$$E = -0.00966\ (MW) + 46.1 \qquad (4)$$

where MW = molecular weight (g/gmol)

Collision rate values for aromatics, alkanes and alkenes can be determined based on their molecular weights. These values for toluene and benzene are 2.85×10^{11} and 2.4×10^{11}, respectively. (Cooper and Alley. 1986. Air pollution control: a design approach. PWS Engineering, Boston, MA.)

T-17. Experimental evidence supports the theory that incineration of a sludge in a multiple hearth incinerator follows first order kinetics with a time time response as described below: (ct)

$$\frac{dC}{dt} = -k\ C(t - d)$$

where
C(t) = the concentration of the reacting compound at time t
k = the reaction rate constant
d = the delay time (sec) = a - b (T)
a = 2.5 s
b = 10^{-3} s/°F

The operating temperature of the incinerator is 1800 °F, the temperature gradient is 1.5 °F/ft, and the reactor length = 40 ft.

Assume a four-hearth incinerator with a rate constant described by the following equation:

$$k = A\ e^{-\frac{E}{RT}}$$

where
A = 3×10^7/s
E = 40 kcal/gmol

a. Calculate the required retention time to achieve 99.99% destruction efficiency. Assume equal reaction times in each hearth.
b. Calculate the total required retention time for 99.99% destruction efficiency in a two hearth incinerator if the reaction rates and individual delays at each hearth are k_1 = 3/s, k_2 = 5/s, d_1 = 1 s, and d_2 = 3 s. Assume equal reaction times in each hearth.

T-18. The gas velocity in a hazardous waste incinerator is 10 ft/s. The temperature at the incineration point in the inlet of the incinerator chamber is 2000 °F, and there is a temperature gradient of 2 °F/ft in the unit. Calculate the minimal length of the incineration chamber for 99.99% destruction efficiency of a compound if its combustion reaction is first order with an expression for k as indicated in Problem T-19 above, E = 50 kcal/gmol, and A = 2 x 10^8/s. (kg)

CHAPTER FIVE

Incinerators

I-1. Since many different liquids are incinerated in most operating incinerators, a blending system is used to prepare feed stock (volume for several hours feed) such that the composition to the incinerator is approximately constant. If the "batch" is not adequately mixed, the composition, and therefore required operating conditions (residence time, temperature and air feed rate), may change. For a two liquid feed, assume that the feed mixture changes from 20% (lbmol basis) A to 40% (lbmol basis) A. The characteristics of Waste A and Waste B are shown below: (flw)

Waste A:
hydrocarbon oil, $C_{30}H_{60}$ containing 0.1% (lbmol basis) hexachlorobenzene.
Waste B:
waste organic, $C_{15}H_{15}Cl_5O_3$ containing 10% (lbmol basis) water

a. Calculate the new air requirement if the incinerator is to operate at 50% excess air under both conditions.
b. What will be the effect on incinerator performance if the air rate is not changed?

I-2. Granular activated carbon with an ash content of 6% (by weight) has been used to adsorb 2,3-dichlorophenol ($C_6H_4OCl_2$) at a loading is 0.31 g/g carbon. The moisture (water) content is 0.1 g/g carbon. This hazardous waste will be fed to a 6 ft x 18 ft rotary kiln followed by an afterburner. To protect the refractory, the kiln temperature is not to exceed 2400°F. For compliance considerations the excess air should be 50%. Assume 5% heat loss. (jmh)

$(-\Delta hc)$ carbon = 14,086 Btu/lb
$(-\Delta hc)$ solvent = 21,032 Btu/lb

a. How many gallons of water must be added to each 100 lb of hazardous waste to control the operating temperature?
b. How many pounds per hour of original waste (before water addition) may be fed to the incinerator if the design criteria call for a heat release rate of 25,000 Btu/h/ft^3?
c. Adding water is one approach to control operating temperature. What other approaches could be used? Would these other approaches be preferable?

I-3. Special fire brick (thermal conductivity = 0.028 Btu/h-ft-°F and an ordinary brick (thermal conductivity = 0.82 Btu/h-ft-°F will be used to enclose an incinerator. The temperature inside the incinerator is 3,000 °F. If the maximum temperature ordinary brick can withstand is 700 °F, what should be the thickness of the special brick? Inside diameter of the incinerator is 10 ft. The outside temperature of the ordinary brick is 100 °F, and its thickness is 4 inch. (ssl)

I-4. Compare the reactor volumes required to decompose diethyl peroxide at 225 °C. The pollutant enters the incinerator at 12.1 L/s and the required destruction efficiency is 99.995%. The kinetic data available include: (jsw)

$R = -kC$ gmol/L-s $k = 38.3/s$ @ 225 °C

a. Carry out the calculations assuming a plug flow model applies.
b. Carry out the calculations assuming a continuously stirred tank reactor (CSTR) model applies. Also perform this calculation with two CSTRs in series.

I-5. An air stream contaminated with 30 ppm chlorobenzene is to be discharged at 40,000 acfm (125 °F and 1 psig) to a hazardous fume incinerator (HFI). The HFI is to be fired with methane and operate at 2000 °F. The residence time is to be 1.8 s. A heat transfer loss of 5% is reasonable. The length to diameter ratio for the HFI is to be 3.0.

Calculate the following: (jmh)

a. The methane flow rate required, scfm.
b. The length and diameter of the HFI.
c. The heat release rate in $Btu/ft^3/hr$.

I-6. A mixture of 1% (by weight) $C_6H_5CHCl_2$ and #6 fuel oil with 3% (by weight) sulfur content is fired in a liquid injection incinerator with 25% EA. (jmh)

a. Calculate the maximum operating temperature (°F) expected for this incinerator. Assume 5% heat loss. The fuel oil analysis is:

Carbon	86.7%
Hydrogen	10.3%
Sulfur	3.0%

b. To what value would the EA need to be changed in order to drop the operating temperature by 300 °F?
c. If the waste injection rate is to be 400 gal/h (2500 lb/h) what length and diameter incinerator would you suggest to be used based on a heat release rate of 25,000 $Btu/h\text{-}ft^3$?

I-7. The solid particles in a fluidized bed incinerator are sand with a diameter of 0.15 mm and a density of 1500 kg/m^3. Air for fluidization is supplied at 1.5 atm and 800 K. Assume the void fraction at fluidization e = 0.56. The viscosity and density of air at 800 K and 1.5 atm are 4.3×10^{-5} Pa-s and 0.66 kg/m^3, respectively. (jhm)

a. Given a cross-sectional area of 0.5 m^2 with 500 kg of solid, calculate the minimum height of the fluidized bed.
b. Calculate the pressure drop at minimum fluidizing conditions.
c. Using the empirical equation given below, calculate the minimum velocity for fluidization.

Reference: Geankoplis, C.J., "Transport Processes: Momentum, Heat and Mass," Allyn and Bacon, Inc 1983.

Equation to find minimum height of fluidized bed:

$$L_1/L_{mf} = (1 - e_{mf})/(1 - e_1)$$

where L_1 = initial height
L_{mf} = height at minimum fluidization
e_{mf} = void fraction at minimum fluidization
e_1 = void fraction at initial conditions taken as zero

Pressure Drop ΔP is given by:

$$\Delta P = L_{mf}\ (1 - e_{mf})\ (\rho_p - \rho)\ g$$

I-7 (Continued).

An empirical equation by Wen, C.Y. and Yu, Y.H., AICHEJ 12, 610 (1966) is also available for the calculation of Reynold's Numbers for minimum fluidization as:

$$N_{Re} = \left[33.7^2 + 0.0408\, D_p^3 \frac{\rho\,(\rho_p - \rho)\, g}{\mu^2} \right]^{1/2} - 33.7$$

where
N_{Re} = Reynold's Number (dimensionless) = $v\, D_p\, \rho/\mu$
v = minimum velocity of fluidization (m/s)
D_p = particle diameter (m)
ρ = gas density (kg/m^3)
ρ_p = particle density (kg/m^3)
μ = gas viscosity (Pa-s)
g = 9.81 m/s^2

For Reynolds Numbers ≈ 0.001 to 4000 this equation gives N_{Re} values within 25%.

I-8 A cellulose waste is incinerated in a fluidized bed incinerator with limestone ($CA(OH)_2$) addition in order to eliminate acid removal at a later stage. The total feed rate is 800 lb/h, while the limestone addition rate is 40 lb/h. The following relation can be assumed between NHV (Btu/lb), temperature (T,.°F) and excess air (EA): (ct)

$$NHV = \frac{0.3\,(T - 60)}{1 - (1 - EA)(7.5 \times 10^{-4})(0.3)(T - 60)}$$

a. Find the mass flow rates of the incinerator products (assume 5% moisture in combustion air).
b. Calculate the operating temperature with 100% EA assuming adiabatic operation.
c. Calculate the required excess air so that the bed temperature (assumed uniform) is 2200 °F.
d. Calculate the total fuel requirements for heating to 2300 °F with 100% EA if the heating value of the fuel is 10,000 Btu/lb.

I-9 A rotary kiln used in the incineration of hazardous waste is slightly inclined to the horizontal. As the cylinder rotates around its axis, the waste charge undergoes mixing and travels toward the lower end of the cylinder. Calculate the residence time for a high density solid (no blowout by gases) when a kiln L ft long and D ft in diameter rotates with N rpm. The slope of the kiln is S ft/ft of length. (ssl)

I-10 Assume that you are burning polychlorinated biphenyls (PCBs) in a rotary kiln incinerator. The principal pollutant, PCB, is mixed with waste fuel oil. Keeping all other variables constant, i.e., excess air, heat loss, and feed rate, what will happen as the percentage of PCB is increased in the waste mixture? (lm)

I-11 A rotary kiln is to be used to destroy a chlorinated herbicide which has recently been banned from agricultural use. From a lab study it has been determined that incineration kinetics follow an approximate first order, with the first order rate constant taking the form: (bmt)

$$k\ (1/s) = A\ e^{-\frac{E_{act}}{RT}}$$

where $A = 3.3 \times 10^7$ (1/s) and E_{act} = 25 kcal/gmol

A stochastic analysis shows that Q_{cg} = 350 scf/lb at 60 °F where Q_{cg} = combustion gas flow rate at 150% EA. A design study is to be performed to determine kiln operating performance. Kiln operating conditions are: maximum superficial velocity = 20 fps, and θ = 2 s. Perform the following calculations at 1700 °F and 2200 °F operating temperatures and with an herbicide feed rate of 1000 lb/h:

a. The actual combustion gas flow rates.
b. k at 1700 °F and 2200 °F.
c. L and D for the incinerator.
d. The destruction efficiency assuming a complete mixed reactor.
e. The destruction efficiency assuming a plug flow reactor.

I-12 A filter press sludge cake from a lime precipitation process treating chromium plating waste is to be dried and recalcinated in a rotary kiln furnace. Estimate the amount of methane required as fuel in lb/hr if 10% EA is to be used. Assume all feeds are at 60 °F. (rfw)

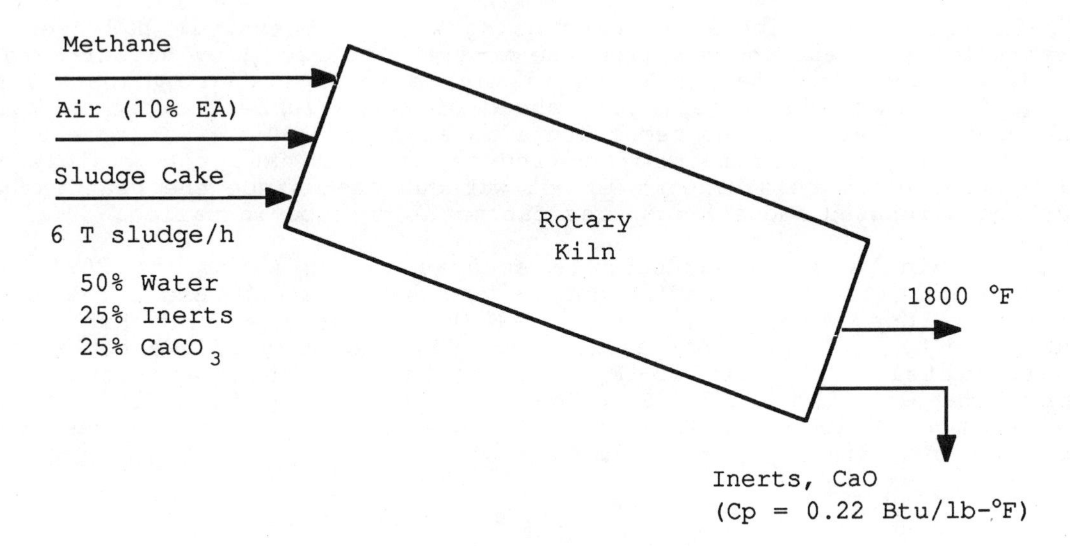

CHAPTER SIX

Waste Heat Boilers/Quenchers

WHB/Q-1. A flue gas at 2000 °F is to be used to generate 100 psia steam (328 °F) in a waste heat, fire tube boiler. The flue gas should be cooled to 380 °F and it can be assumed that feed water is available at its boiling point. How many 1 inch tubes 20 ft in length are necessary in the boiler to be able to cool 50,000 lb/h of gas? The gas may be assumed to be 80% N_2, 10% O_2, and 10 % CO_2. Calculate the number of tubes by: (jhm)

a. Assuming the flue gas film coefficient is controlling such that the overall heat transfer coefficient, U (Btu/h-ft^2-ft-°F) may be calculated from $U = 1986/N^{0.8}$, where N = number of tubes.
b. Using the nomograph for waste heat boiler performance evaluation given in Appendix M. Use C_p data from Appendix C to solve the problem.

$$q = w_h C_p (T_{H1} - T_{H2}) = U\ A\ \Delta T_{ln\ mean}$$

where
W_h = gas flow rate (lb/h)
C_p = heat capacity of the flue gas (Btu/lb-°F)
T_{H1} = inlet flue gas temperature (°F)
T_{H2} = outlet flue gas temperature (°F)
T_s = steam temperature (°F)
U = overall heat transfer coefficient (Btu/hr-ft^2-ft-°F) based on inside area
$\Delta T_{ln\ mean}$ = $[(T_{H1} - T_s) - (T_{H2} - T_s)]/\{\ln[(T_{H1} - T_s)/(T_{H1} - T_s)]\}$
A = total heat transfer area inside the tubes = πDNL
D = inside diameter of tube (ft)
L = total tube length (ft)

WHB/Q-2. In many incinerator units using wet scrubbers, it has been found desirable to reheat the gas from the scrubber to 220 °F or higher before it is discharged from the stack. This may be accomplished by installing an economizer (heat exchanger) just ahead of the scrubber, or at a point in the process where the gas temperature is at least 350 °F. This reheat step can reduce the stack height and reduce or eliminate the visible steam plume. However, this benefit is not without cost since the heat exchanger must be purchased and a larger gas fan motor must be installed.

The gas leaving a given scrubber is saturated with water at 120 °F. The overall heat transfer coefficient is assumed to be 25 Btu/h-ft^2-°F. The pressure loss across the additional duct plus the heat exchanger is estimated to be 0.1 psi for each side. Heat capacity of the hot stream is approximately 0.25 Btu/lb-°F, while that of the saturated gas is approximately 0.30 Btu/lb-°F. The hot gas density is 0.05 lb/ft^2, and contains 0.04 lb water/lb dry gas. Use the following expressions to estimate heat exchanger costs: (flw)

$$C \approx 40{,}000\ A^{-0.44}\ e^{0.067\ (\ln A)^2}$$

where C = heat exchanger cost ($)
A = heat exchanger area (ft^2)
Installed $ = 3 C

a. Estimate the cost of a heat exchanger for this service if 10,000 acfm of gas at 400 °F and 0.1 psig are available.
b. Estimate the power cost of this addition if fan efficiency is 60% and electric power costs $0.10/kwh.

WHB/Q-3. A water quench is used to cool 40,000 acfm of combustion gases containing 30% water vapor, 8% CO_2, 5% O_2, and 57% N_2 (volume basis) from 2,000 °F to the adiabatic saturation temperature. Assume an average heat capacity of the combustion gases of 8.67 Btu/lbmol-°F. Water at 70 °F is used. Determine the following: (heh)

a. The exit temperature of the quenched gases.
b. The amount of water theoretically required for quenching.
c. The volumetric flow rate of the exit gases.

WHB/Q-4. Incineration of a hazardous waste at 1800 °F produces 10,000 lb/h of dry gas and 1500 lb/h of water. Heat is to be recovered in a waste heat boiler by generating steam at 200 psia. Water is fed to the boiler at 218 °F. Some water is lost from the boiler as blow-down amounting to 1/20 of the feed water and the remaining 19/20 of the feed is converted to steam. Ninety percent of the steam generated is fed to a turbine and ten percent is mixed with the condensate return from the turbine and make-up water which is fed to the boiler as feed water. Determine the amounts of: steam produced, boiler feed water, make-up water, steam added to the feed water, and steam condensate from the turbine. (ct)

Data: The ΔT across the tubes of the waste heat boiler is 140 °F, condensate return is at 180 °F and feed make-up water is at 70 °F. Radiation losses from the boiler amount to 1% of total heat input.

CHAPTER SEVEN

Air Pollution Control Equipment

APCE-1. List the advantages and disadvantages of using electrostatic precipitators, fabric filters, venturi scrubbers, wet ionization precipitators, spray dryers, and wet scrubbers to separate and remove pollutants from flue gases that are generated by the combustion of hazardous wastes. (mjr)

APCE-2. A conventional Lapple cyclone is used to remove particulates from a flue gas upstream of a horizontal secondary combustion chamber. The actual particulate concentration in the flue gas is 5.0 gr/ft^3 at 1600 °F and 1 atm. The particle specific gravity is 2.9. The actual gas flow rate is 4520 acfm. The diameter of the cyclone is 3.0 ft. What is the particulate concentration at the outlet of the cyclone? (mjr)

The height and width of the inlet duct to the cyclone are 0.5 and o.25, respectively, of the cyclone diameter. The efficiency of teh cycleone may be estimated using the following equation:

$$E = 1.0/[1.0 + (d_{p,50}/d_p)^2]$$

where dp = particle diameter (μm)
$d_{p,50}$ = particle cut diameter (μm)

The particle cut diameter is calculated from Lapple's equation (with consistent units):

$$d_{p,50} = \left(\frac{9\ \mu\ W}{2\ \pi\ N\ V_i\ (\rho_p - \rho_g)} \right)^{0.5}$$

where μ = gas viscosity
W = cyclone inlet width
N = number of turns of the gas stream within the cyclone
V_i = inlet gas velocity
ρ_p, ρ_g = particle and gas density

The following particle size distribution should be used for solving the problem:

Particle Size Range (μm)	Mean Diameter (μm)	Mass Fraction
0 - 10	5	0.25
10 - 40	25	0.50
40 - 80	60	0.25

APCE-3. A fly ash laden gas stream is to be cleaned by a venturi scrubber. The fly ash has a particle density of 43.7 lb/ft^3 and k = 0.2 ft^3/1000 gal. The throat velocity is 272 ft/s and the gas viscosity is 1.5 x 10^{-5} lb/ft-s. Due to the concern over emissions of fine particulates associated with metals, the L/G ratio and pressure drop required to achieve 90% and 80% removal of the 0.5 μm and 0.3 μm particles, respectively, must be determined. (jsw)

The following equation can be used for the calculation of mean droplet diameter:

$$d_{mean} = 16,400/V_t + 1.45\ R^{1.5}$$

APCE-3 (Continued).

where Vt = venturi throat velocity (ft/s)
R = liquid to gas ratio (gal/1000 ft^3)

Removal efficiency for the scrubber can be described as:

$$E = 1 - \exp(-kRA^{0.5})$$

where k = 0.2 ft^2 gas/1000 gal
A = $(C\ d_p^2\ \rho_p\ V_t)/(9\ \mu_g\ d_{mean})$
C = 1.325 @ 70 °F for 0.5 µm particles, 1.5 @ 70 °F for 0.3 µm particles
d_p = particle diameter (µm)
ρ_p = particle density (lb/ft)
μ_g = gas viscosity (lb/ft-s)

Finally, the pressure drop accross the venturi scrubber, in inches of water, can be expressed as:

$$\Delta P = 5 \times 10^{-5}\ V_t^2\ R$$

APCE-4. Granular activated carbon to be used for fugitive emissions control on a waste tank produced the breakthrough curve data in Table 1 for trichloroethylene (TCE) in a bench scale experiment with air at 20 °C containing 5000 ppmv of TCE. A carbon bed of 100 g was used. (seh)

a. Determine by plotting outlet concentration versus total sample volume treated what the breakthrough loading (g TCE/g carbon) would be for 99% TCE removal.
b. Waste determined to be 28 wt% TCE is to be stored in a cylindrical tank 3 m in diameter and 4 m high (effective dimensions). Upon emptying, the tank may be assumed to contain vapor concentrations at TCE's saturation vapor density at 20 °C. When the tank is re-filled the effective tank volume, saturated with TCE, will be displaced through the tank vent to the activated carbon trap. If the trap contains 6 kg of carbon how many drain/fill cycles can be performed before the carbon must be replaced or regenerated?
c. Use the breakthrough curve to estimate the mass of TCE discharged from operating one full cycle beyond this limit.
d. If the TCE contaminated waste was incinerated to 99.99% DRE, what fraction would the TCE emission, calculated in Part c, be of the total emission from incinerating one complete tank of the TCE waste? The waste has a bulk density of 1.13 kg/L at 20 °C.

Table 1. Granular activated carbon break-through data for TCE fugitive emission control

Total Sample Volume Treated (L)	Outlet TCE concentration (ppmv)
0	1
25	3
50	6
75	10
100	21
150	125
175	182
200	223

APCE-5. 50,000 scfm of gas flows through a bag house with an average filtration velocity (v) of 10 ft/min. The pressure drop (inches of water) is given as:

$$\Delta P = 0.2\ v + 5\ W_i\ v^2\ t$$

where W_i = dust loading (lb/ft^3)
t = cumulative filtration time (min)

The fan can maintain this volumetric flow rate up to a pressure drop of 8 inches of water. Laboratory experiments were conducted to determine the coefficients, ψ (250 1/ft) and ϕ (0.152 1/min) in the following particle removal equation:

$$E = 1 - e^{-(\varphi L - \phi t)}$$

The fabric thickness (L) is 0.01 inches. Calculate the average concentration of particulates emitted and the pressure drop at the end of a cleaning cycle. Use dust loadings of 2 and 5 gr/ft^3. (jsw)

APCE-6. Flue gas from a secondary combustor is passed through a spray tower followed by a bag house to remove particulates. The flue gas from the bag house has 12% water vapor by volume and its mass flow rate is 5,000 lb/h. This gas at 200 °F is counter-currently contacted by water in a packed tower to remove moisture by cooling. Assume the flue gas has the following composition by volume:

$$CO_2=10\%,\ H_2O=12\%,\ N_2=73\%,\ O_2=5\%.$$

If the flue gas leaves at 100 °F from the cooling tower, calculate the amount of moisture removed from the flue gas in lb/h. (ssl)

APCE-7. Estimate the lime feed rate (dry basis) and ash generation rate (dry basis) for the system and conditions shown below using double the stoichiometric lime requirement. The expected performance results are: (rfw)

HCl = 99.9% removal
CO_2 = 1% absorbed
SO_2 = 75% removal
Particulates = 0.01 gr/scf

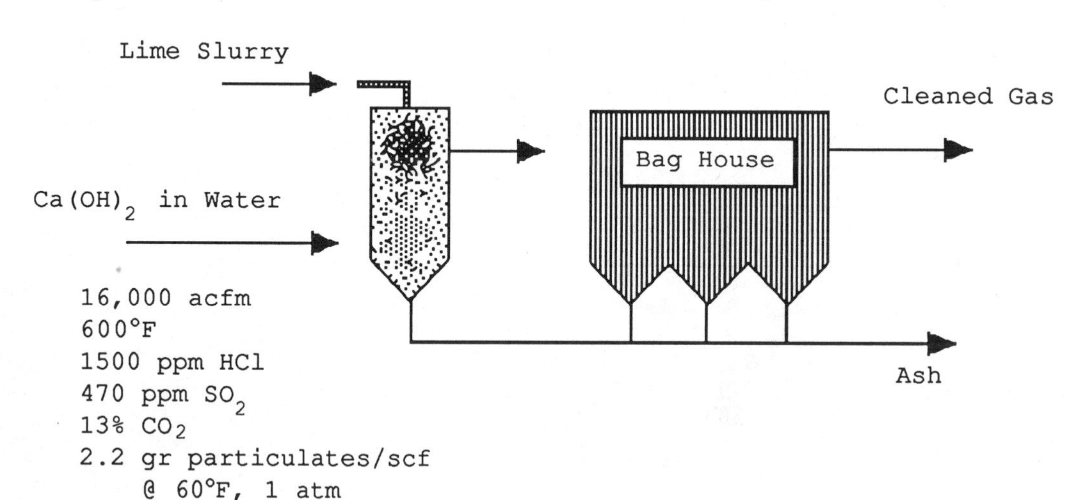

APCE-7 (Continued).

Key neutralization equations are given below:

$$HCl + 1/2\ Ca(OH)_2 \rightarrow 1/2\ CaCl_2 + H_2O$$

$$SO_2 + Ca(OH)_2 \rightarrow CaSO_3 + H_2O$$

$$CO_2 + Ca(OH)_2 \rightarrow CaCO_3 + H_2O$$

APCE-8. A two compartment reverse air fabric filter is used to separate and remove particulate material that is generated by the combustion of hazardous waste. The particulate loading is 5.0 gr/acf, and the superficial gas velocity in the fabric filter is 2.0 ft/min. Experimental results describing pilot plant data of filter drag and particulate mass loading are presented below:

Filter Drag [Pa-min/m]	Particulate Mass Loading [g/m^2]
153	0
37	22
504	45
612	90
693	135
990	270

Only one compartment is on-line at all times. Maximum pressure drop of the fabric filter is 2000 Pa. What is the maximum filtration time before a compartment is taken off-line for cleaning? (mjr)

The maximum pressure drop of a fabric filter, S, can be expressed using the following equations:

$$S = K_1 + K_2\ W = K_1 + K_2\ C_p\ V_g\ t = \Delta P/V_g$$

where K_1, K_2 = residual cloth filter drag coefficients estimated from a regression of filter drag versus particulate mass loading
K_1 = intercept of regression equation (Pa-min/m)
K_2 = slope of regression equation (Pa-min-m/g)
W = particulate loading rate (g/m2)
C_p = particulate concentration (g/m3)
V_g = superficial gas velocity (m/min)
t = cumulative filtering time (min)

APCE-9. An incinerator emits 40,000 lb gases/h containing 200 lb HCl/h and 8,000 lb/h water. The molecular weight of the flue gas is 28. Ninety-nine % of the HCl must be controlled. A vertical scrubber packed with telleretter is to be used in counter-current operation. Determine the following: (heh)

a. the size of the scrubber operating at 275 °F.
b. the scrubber superficial gas velocity.
c. the amount of recycle liquid required at a liquid to gas ratio of 25 gal/1000 acf.
d. the amount of HCl removed in lb/h.

Use an approximate solution approach were tower area = 0.2 (mass gas flow rate)/(hot gas density) and tower height = 2 (tower diameter)

APCE-10. A gas stream containing 2.0% by volume of HCl is to be scrubbed counter-currently. The scrubber is to remove 99.5% of the HCl with a liquid flow 150% above the minimum. Henry's Law constant for HCl = 0.2. (sb)

a. Calculate the molar liquid/gas ratio if clean water is used in the scrubber.
b. Calculate the molar liquid/gas ratio if 50% of the used scrubbing water is recycled with clean water.

APCE-11. A packed column was designed to remove 99% of the HCl from a gas stream containing 2% HCl. It has been discovered that inlet HCl concentrations may reach 5%. Will it be possible to use the column previously designed to reduce the concentrations to the same level as planned for the 2% HCl stream? The original column was designed with an L/G ratio equal to 500% of the minimum. What will be the relationship between L/G_{actual} and the minimum L/G ratio for the new conditions? (jmm)

<u>Operating Data and Assumptions</u>
Liquid flow rate = 3690 lb/h
Gas mass flow rate = 5000 lb/h
Scrubbing liquid = pure H_2O
Packing = 1 in Raschig rings
H_{OG} = 2.5 ft at the operating conditions
Henry's Law constant = 0.2
Column height = 15 ft
Column Diameter = 2.6 ft

Use the following equations for the packed scrubber design:

Overall component balance to remove 99% of HCl with 1 % remaining:

$$y_2 = (0.01\ y_1)/[(1 - y_1) + (0.01\ y_1)]$$

$$Z = N_{OG}\ H_{OG}$$

$$N_{OG} = \ln\{[(y_1 - m\ x_2)/(y_2 - m\ x_2)]\ (1 - 1/A) + 1/A\}/(a - 1/A)$$

$$A = L/(m\ G)$$

where
y_2 = mol fraction of gas exiting column
y_1 = mol fraction of gas entering column
x_2 = mol fraction of liquid exiting column
x_1 = mol fraction of liquid entering column
x_1^* = equilibrium mol fraction in liquid, $x_1^* = y_1/m$
Z = height of column (ft)
N_{OG} = number of transfer units (dimensionless)
H_{OG} = height of a transfer = 2.5 ft
m = slope of equilibrium line (mol fraction/mol fraction)
L = liquid flow rate (lb/h)
G = gas flow rate (lb/h)

APCE-12. A packed column is to be used to absorb HCl from a flue gas stream. In order to meet EPA standards, HCl emissions must be less than 4 lb/h. Calculate the tower height and column diameter for the following conditions: (ssl)

Gas flow rate = 4,000 lb/h
HCl concentration in inlet gas stream = 5% by volume
packing type = 1 in Raschig rings
HOG of the tower = 2.5 ft
Henry's Law constant = 0.2
density of the flue gas = 0.075 lb/ft^3
Pure water at 70 °F will be used as a scrubbing liquid.
The tower operates at 60% of the flooding velocity (flooding velocity is 500 lb gas/h-ft^2), and the water flow rate is two times the minimum.

APCE-13. For the packed bed absorber shown, calculate: (rfw)

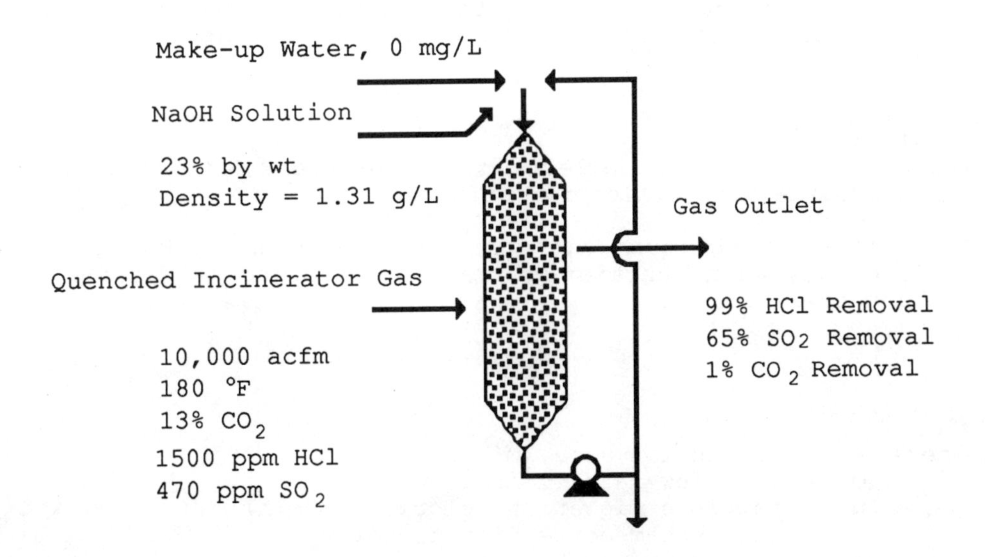

a. The mass flow rate of NaOH required in gal/h.
b. The purge flow rate in gal/h. Assume that the purge flow (blow-down) is 5% TDS, with a specific gravity = 1.0.

APCE-14. Combustion of a hazardous waste produces 10,000 acfm of flue gas at 750 °F and 1 atm. HCl and SO_2 concentrations are 20,000 ppm and 350 ppm, respectively. HCl must be controlled to 99% collection efficiency or 4 lb/h. SO_2 emissions are controlled at 70% removal efficiency. A spray dryer is used to control the HCl and SO_2 emissions. $Ca(OH)_2$ is the sorbent that will react with HCl and SO_2 to form $CaCl_2$ and $CaSO_4$, respectively. Assume the necessary stoichiometric ratios of the $Ca(OH)_2$ to $CaCl_2$ and $CaSO_4$ (assuming complete chemical reaction of the SO_2 and HCl with $Ca(OH)_2$) are 1.1 and 1.3 to achieve the HCl and SO_2 control requirements, respectively. (mjr)

a. What is the required feed rate of $Ca(OH)_2$?
b. What is the total mass production rate of solids if the water content of the spray dryer's by-product is 10% by mass?

APCE-15. An incinerator burns mercury contaminated waste. The waste material has an ash content of 1%. The solid waste feed rate is 1000 lb/h and the gas flow rate is 20,000 dscfm. The average mercury content of the particulate matter is 2.42 μg/g, while the vapor concentration is 0.3 mg/dscm. For the case when the incinerator meets the particulate standard of 0.08 gr/dscf (0.1832 g/dscm) with a 99.5% efficient ESP, calculate: (kg)

a. The amount of mercury (lb/d) bound to the fly ash which is captured in the ESP.
b. The amount of mercury (lb/d) leaving the stack as a vapor and with the fly ash.

APCE-16. A hazardous waste incinerator emits a particulate loading of 5.0 gr/dscf corrected to 50% excess air upstream of an electrostatic precipitator. Thirty percent of the particulate mass is 0.3 μm in diameter. The balance of the particulate material is 5.0 μm in diameter. The flow rate of the flue gas is 5,000 m^3/min at 177 °C and 1 atm. The dielectric constant of the particulate material is 5. The ESP's field strength is 300 kV/m. Assume Cunningham correction factors of unity. (mjr)

a. What is the ESP's minimum required particulate mass collection efficiency?
b. What is the minimum collection area of the ESP to achieve the required particulate mass collection efficiency?

Use the following equations for the prediction of ESP particle removal efficiency as a function of system operating conditions and particle characteristics.

$$E = 1 - e^{-\frac{A\,u}{Q}}$$

where
- A = collection area (m^2)
- Q = gas flow rate (m^3/min)
- u = particle migration velocity (m/min) = $(C\ d_p\ \varepsilon_o\ K\ E^2)/(3\ \mu_g)$
- C = Cunningham correction factor = 1.0
- dp = particle diameter (m)
- ε_o = particle resistivity = 8.85×10^{-2} c/v-m
- K = constant = 5
- E = electric field strength (v/m)
- μ_g = gas viscosity = 2.48×10^{-5} kg/m-s @ 177 °C

CHAPTER EIGHT

Risk Analysis

RA-1. Many operational parameters at a hazardous waste incinerator are continuously monitored. Values outside of desired ranges may signal impending problems with: a) contaminant destruction and removal efficiency, b) equipment integrity, and/or c) worker safety. For each of the following indicators, describe concerns for one or more of the problem categories listed above. (seh)

a. high CO in stack gas
b. low temperature in the secondary combustion chamber
c. high combustion gas flow rate
d. low scrubber water flow rate
e. low quench water flow rate
f. high pressure in primary or secondary combustion chamber
g. high temperature in primary or secondary combustion chamber
h. loss of flame in primary or secondary combustion chamber

RA-2. A 10,000 gal/d feed tank at a liquid incinerator is filled and emptied everyday. During filling (@ 75°F) the tank is vented to the atmosphere. Which emits more hydrocarbons, the tank vent or the incinerator? (rfw)

Assumptions:

- The waste is contaminated Benzene
  Vapor Pressure.= 95 mm Hg @ 75°F
  Specific Gravity = 0.87 g/cm^3
- Tank contents @ 75°F
- Tank vapor at equilibrium with liquid
- The incinerator achieves 99.99% DRE

RA-3. Risk may be logically considered to consist of:

- a scenario
- a probability of the scenario occurring, and
- a set of consequences which arise if the scenario does occur.

Consider a rotary kiln incinerator which is receiving drummed wastes with highly variable heating values. The process train consists of the following: the feed system, the primary combustion chamber, a secondary combustion chamber designed to operate at 1200°C, an emergency vent stack, a quencher, a packed tower scrubber and two ionizing wet scrubbers in series connected to an induced draft fan and a 20 m stack. (seh)

a. Propose a scenario which may lead to a discharge of POHCs which is substantially higher than the permitted DRE.
b. Suggest a logical approach for estimating the probability of that scenario using a process flow sheet, the interdependence of equipment and/or operator intervention, and estimates of the probability of individual events occurring.
c. Describe how you would approach estimating the consequences of your scenario based upon a typical hazardous waste feed to the system.
d. For your example scenario, how could risk be reduced by dealing with one or more of the three components of risk?

(The answer(s) to this problem are open-ended. The reader should think about the entire incineration process, consider how upset conditions might be anticipated and consider how the process might be improved to reduce the overall risk of undesirable consequences.)

RA-3 (Continued).

References:

Guidelines for Hazard Evaluation Procedures. A.I.Ch.E. Center for Chemical Process Safety. Publ.No. LC 85-236087. ISBN 0-8169-0399-9.
Guidelines for Chemical Process Quantitative Risk Assessment Procedures. A.I.Ch.E. Center for Chemical Process Safety Publ. In press.

RA-4. Plume modeling of emissions from stacks is usually performed using some version of the diffusion equation based upon a Gaussian distribution. A simplified form of this equation which can be used to estimate the maximum ground level concentration from a continuous point source is given as:

$$C(x,y,0;H) = \frac{Q}{2\pi\,\sigma_y \sigma_z\, u} \exp\left(-\frac{1}{2}\left(\frac{Y}{\sigma_y}\right)^2\right) \exp\left(-\frac{1}{2}\left(\frac{H}{\sigma_z}\right)^2\right)$$

where C = contaminant concentration (mg/m^3)
x = downwind distance along plume centerline (km)
y = crosswind distance (km)
H = effective stack height, actual stack height + ΔH (m)
Q = contaminant emission rate (mg/s)
σ_y, σ_z = dispersion coefficients in the x and y directions, respectively, (m)
u = mean wind speed (m/s)

The plume rise necessary to calculate the effective plume height can be estimated from the Holland equation:

$$\Delta H = \frac{v\,d}{u}\left[1.5 + 0.0056\,P\,d\left[\frac{T_s - T_a}{T_s}\right]\right]$$

where ΔH = plume rise above stack exit (m)
v = stack gas exit velocity (m/s)
d = inside stack diameter (m)
P = atmospheric pressure (kPa)
T_s = stack gas exit temperature (K)
T_a = ambient temperature (K)

Evaluate the maximum ground level concentration of suspended fine particulate matter for stable meteorological conditions for a hazardous waste incinerator which will achieve a discharge concentration of 20 mg/m^3 for a discharge flow of 20,000 m^3/h through a 20 m stack with a diameter of 0.75 m. The stack gas exit velocity is 13 m/s at a temperature of 42°C. Ambient conditions are: temperature = 15°C, wind speed = 2 m/s, pressure = 101 kPa, and stability category F. Obtain the σ_y and σ_z from Appendix J. (seh)

a. Plot ground level concentrations between 0.5 and 10 km downwind and determine the location of maximum particulate concentration for the specified conditions.
b. Compare this concentration with the following data measured for suspended particulates from cigarette smoke (Sterling et al. 1982. J. Air Poll. Control Assoc. 32, 250-259.)

RA-4 (Continued).

Tavern - 0.99 mg/m^3
Church bingo - 0.28 mg/m^3 (20 smokers)
Restaurant - 0.086 mg/m^3 (one smoker in 781 m^3)

In which of these locations is it safest to be? Justify your answer.

RA-5. Estimate the impact (dosage and maximum concentration) on the area surrounding a waste incineration unit if the unit loses electric power such that all air and waste feed is stopped, and the emergency vent is activated. Assume that uncombusted waste equivalent to 10 s of feed is released as a "puff." The emergency vent is 10 m above the ground and the waste feed rate is 7200 lb/h. Weather conditions are: wind speed = 10 mi/h, overcast conditions (neutral stability). Ignore plume rise, and obtain σ_y and σ_z from Appendix J. Utilize the following equation for the calculation of ground level concentration for a puff of total mass release, Q_r, at a distance x (m), for a wind speed u (m/s), and an effective height H (m). (flw)

$$C(x,0,0;H) = \frac{2\,Q_r}{(2\pi)^{3/2}\,\sigma_x\,\sigma_y\,\sigma_z}\exp\left[-\frac{1}{2}\left(\frac{x-u\,t}{\sigma_x}\right)^2\right]\exp\left[-\frac{1}{2}\left(\frac{H}{\sigma_z}\right)^2\right]\exp\left[-\frac{1}{2}\left(\frac{y}{\sigma_y}\right)^2\right]$$

a. Evaluate the location of the maximum ground level concentration of released waste feed. Include calculations at downwind distances of 100 m, 1000 m and 4000 m.
b. Estimate the dosage (30 min averaging time) of this uncombusted waste feed at the same locations using the following equation and data which yield the average to peak concentrations for a pollutant based on the number of σ_x values which have passed a given receptor point, x, over an averaging time, t, assuming uniform wind speed during this averaging period.

For the number of σ_x values, M, greater than 10, average to peak concentrations are approximated as:

Concentration Average/Concentration Peak = 2.5067/M

For the number of σ_x values, M, less than 10, average to peak concentrations are:

Number of σ Values	Average/Peak
0.001	1.000
0.2	0.995
0.4	0.991
0.6	0.982
1.0	0.958
2.0	0.853
4.0	0.722
6.0	0.416
8.0	0.312
10.0	0.250

RA-6. Surface deposition and soil concentrations of particulate metals released by a hazardous waste incinerator are to be determined as part of an exposure assessment being conducted. To estimate the settling velocity of the metal particulate material, the following equation for terminal velocity, v_t, is to be used:

$$v_t = \frac{D_p^2 \rho_p g}{18 \mu}$$

where D_p = particle diameter (cm)
ρ_p = density of air (g/cm^3)
g = gravitational acceleration, 980.7 cm/s^2
μ = air viscosity (g/cm-s)

Assume that the incinerator emits the legal maximum particulate concentration as metals, and that it operates at 50% excess air with a stack gas flow of 10,000 dscfm. Also assume that the particles are emitted from an effective height of 200 ft, and that they have a density of 2.5 g/cm^3.

Use the gaussian dispersion equation for the estimation of downwind ambient particulate concentrations. To account for particle settling, a "tilted" plume model may be used, in which H is decreased to account for the vertical motion of the particles over a travel distance x. This model takes the form:

$$C(x,y,0;H) = \frac{Q}{2 \pi u \sigma_y \sigma_z} \exp\left[-\frac{1}{2}\left(\frac{y}{\sigma_y}\right)^2\right] \exp\left[-\frac{1}{2}\left(\frac{H - v_t (t)}{\sigma_z}\right)^2\right]$$

The following expressions may be used for the estimation of dispersion coefficients σ_y and σ_z for stability class D (neutral stability), with x in meters:

$$\sigma_y = 0.32\, x^{0.78} \qquad \sigma_z = 0.22\, x^{0.78}$$

Using the data presented above under conditions of neutral stability with a 5 mi/h wind, determine the following for particles of 0.5, 5, and 50 μm diameter: (sb)

a. Emission rate in lb/h and g/s.
b. Settling velocity of the particles in m/s.
c. Distance at which the centerline of the tilted plume would touch the ground surface in mi.
d. Ambient concentration at the distance determined in Part c in $\mu g/m^3$.
e. Surface flux due to settling at 0.25 mi downwind of the stack in g/m^2-s.
f. Assuming that the meterological conditions described above apply 5% of the time, that the facility operates for 30 years, and that the deposited particles become evenly mixed in the top 10 cm of the soil, calculate the metal concentration in the upper soil in ppm. Use a soil density = 2000 kg/m^3.
g. The annual probability of an individual dying from cancer from metal deposition over a 30 year period. The analysis should be based on the following:

RA-6 (Continued).

1. The probability that a food crop will be planted on the contaminated soil in any given year = 0.25.
2. The probability that a crop will produce a marketable yield = 0.8.
3. The probability that the marketable yield is contaminated will vary with the metal deposition rate and predicted soil concentration as follows:

Metal Soil Concentration (ppm)	Probability of Contamination (independent of D_p)
0.013	0.0001
1.3	0.001
185	0.1

4. The probability of human consumption of contaminated crop initiating a carcinogenic response = 0.005.

h. Discuss the limitations of this analysis and ways to improve it.

CHAPTER NINE

Monitoring

M-1. Determine the minimum volume of flue gas to be sampled to detect 99.99% DRE from an incinerator burning hazardous wastes containing 1% hexachlorobenzene. Also calculate the sampling time if the sample is collected at 1.0 L/min. The waste feed rate is 5000 lb/h and the stack gas flow rate is 427,000 scfm.

Hexachlorobenzene is the POHC for which monitoring is to be conducted. Its analytical detection limit is 10 µg/L. Assume that the sample is to be concentrated in 25 mL solvent. (kg)

M-2. Volatile organic contaminants (those with boiling points > 100°C) are measured in trial burns using an EPA protocol known as the volatile organics sampling train (VOST) method. This procedure involves trapping the volatile organics on a Tenax™ sorbent trap. Analysts must be concerned with both minimum and maximum levels of contaminant to insure adequate quantities for detection while avoiding contaminant breakthrough. A second sorbent tube in series is used to detect any breakthrough occurring during testing. Designated principal organic hazardous constituents (POHCs) must be recovered by the VOST protocol to demonstrate the required destruction and removal efficiency (DRE) for permit approval.

A trial burn is being designed for a waste feed containing three POHCs which have VOST protocol requirements for recovery in the optimum ranges shown below:

	POHC		
Mass Recovered	Carbon tetrachloride	Trichloro-ethylene	Dichloro-methane
minimum	200 ng	250 ng	500 ng
maximum	4000 ng	5000 ng	5000 ng

Samples would normally be collected at a rate of 1.0 L/min for 20 min. (seh)

a. Given an incinerator stack gas flow rate of 500 dscm/min, what is the minimum feed rate of each POHC (kg/h) necessary, with a safety factor of 10, to demonstrate 99.99% DRE for each POHC?
b. What would be the impact on the success of this sampling effort for each POHC if these calculated feed rates were used, but the incinerator achieved only 99.98% DRE?

M-3. POHCs are monitored using the VOST and semi-volatile organic sampling trains (Semi-VOST). The VOST method is intended for POHCs with boiling points from 30 to 100°C , while the Semi-VOST is designed for POHCs with boiling points > 100°C. (seh)

a. For the following POHCs determine whether you would expect them to be detected by the VOST or Semi-VOST method:

carbon tetrachloride	heptachlor epoxide
hexachlorobenzene	chlordane
decachlorobiphenyl	pentachlorophenol
1,1,1-trichloroethane	kepone
hexachlorobutadiene	tetrachloroethylene

b. Will the distribution of compounds between methods be absolute, i.e., all or nothing in one method or another? Why or why not?

M-3 (Continued).

c. These protocols allow for an accuracy of ± 50% in final POHC concentrations. If the true value for a given POHC was just small enough to produce a DRE of 99.99%, what is the tolerance of measured DRE which could arise for a single determination based upon the allowable analytical tolerance? Assume a fixed mass flow into the incinerator of 1.0 kg/h.

d. The EPA proposes that the difficulty of incinerating individual POHCs can be predicted based upon the heat of combustion value for each POHC. Using this rationale, rank the 10 POHCs from Part a in order of most to least difficult to incinerate.

e. Comment on the suitability of using a thermodynamic parameter, such as heat of combustion, for predicting incinerability under reaction conditions with reaction times of only ≈ 2 s. Can you propose an alternative concept which might be more suitable if data were available?

PART 2

PROBLEM SOLUTIONS

This part contains the solutions to the problems presented in Part 1. Each problem solution is identified with the same letters and number used for the corresponding problem statement in Part 1. For Example, the solution to Problem RA-5 in Chapter 8 of Part 1 is identified as Solution RA-5 and can be found in Chapter 8 of Part 2.

CHAPTER ONE

Basic Concepts

BC-1 Solution.

a. 453.6 g/lb x 6.02 x 10^{23} particles/gmol = **2.73 x 10^{26} particles/lbmol of any substance**

b. 453.6 gmol/lbmol x 22.4 L/gmol = **10160 L/lbmol**
10160 L/lbmol x 0.0353 ft^3/L = **359 ft^3/lbmol at 0 °C**
359 ft^3/lbmol x 288.5 K/273 K = **379 ft^3/lbmol at 60 °F**

c. 0.0821 L-atm/gmol-K x 0.0353 ft^3/L x 14.7 psi/atm x 453.6 gmol/lbmol x 0.555 K/°R
= **10.73 psia-ft^3/lbmol-°R**

d. Both absolute temperature scales (Rankine and Kelvin) begin at absolute zero, but temperature in K is smaller than °R by a factor of 5/9. Thus, 2460 °R x 5/9 = **1367 K.**

BC-2 Solution.

From the humidity diagram, Appendix A, H= 0.0805 lb H_2O/lb air at the dry bulb temperature of 180 °F and wet bulb temperature of 125 °F.

a. At humidity H=0.081 lb/lb on the chart, move to the saturation curve and read the dew point ≈ **119.3 °F.**

b. Flow rate = air rate + water rate;
w = flow rate = 10,000 lb/h + 10,000 lb/h (0.0805 lb H_2O/lb air)
w = **10,805 lb/h.**

c. Volumetric flow rate: Since MW of flue gas = 30 lb/lbmol, then mole fraction air in flue gas =

$$\frac{\frac{10,000}{30}}{\frac{10,000}{30} + \frac{10,000\ (0.0805)}{18}} = 0.88$$

mole fraction water vapor =

$$\frac{\frac{10,000\ (0.0805)}{18}}{\frac{10,000}{30} + \frac{10,000\ (0.0805)}{18}} = 0.12$$

Check Σy_i = 1.0 = 0.88 + 0.12 = 1.00 √

Average molecular weight = 0.88 x 30 + 0.12 x 18 = **28.6 lb/lbmol**

Volumetric flow rate on basis of 1 h: w(wet gas) = **10,805 lb/h,**
n = (10,805 lb/h)/(28.6 lb/lbmol) = **378 lbmol/h**

V = nRT/P = (378 lbmol/h x 0.73 atm-ft^3/lbmol-°R) x (460+180 °R)/1 atm
= **1.77 x 10^5 ft^3/h.**

BC-3 Solution.

If the vapor phase is assumed to be ideal, Raoult's Law applies:

$$y_i P_T = P_i = P_i' x_i$$

where xi is the mole fraction of the compound of interest in the liquid phase.

a. % compound by volume in gas phase = $y_{compound} = P_i' x_i/P_T$
toluene: (16 mm Hg/760 mm Hg) 0.6 = 0.013 = 1.3% by volume
benzene: (60 mm Hg/760 mm Hg) 0.4 = 0.032 = 3.2% by volume
Total = 4.5% by volume

b. From Appendix G, the lower flammability limit for toluene is 1.27% by volume, while the lower flammability limit for benzene is 1.4%. Therefore, **the vapor is flammable.**

c. From Appendix G, the NHV of toluene = 4284 Btu/ft^3, while the NHV of benzene = 3601 Btu/ft^3, therefore, the total NHV is:
toluene: 0.013 x 4284 Btu/ft^3 = 56 Btu/ft^3
benzene: 0.032 x 3601 Btu/ft^3 = 115 Btu/ft^3
Total = 170 Btu/ft^3

d. Stoichiometric air required for complete combustion is based on data from Appendix G indicating for toluene, an air requirement of 42.88 ft^3 air/ft^3 toluene, and for benzene an air requirement of 35.73 ft^3 air/ft^3 benzene.
Toluene: 42.88 ft^3/ft^3 x 0.013 = 0.56 ft^3/ft^3
Benzene: 35.73 ft^3/ft^3 x 0.032 = 1.14 ft^3/ft^3
Total = 1.70 ft^3/ft^3

Air present in the gas phase = (1-0.045 volume fraction benzene + toluene)
Air fraction = 0.955

Therefore, **the mixture is fuel "rich" and requires supplemental air for complete combustion.**

BC-4 Solution.

a. $C_2H_6O_2 + 5/2\ O_2 \rightarrow 2\ CO_2 + 3\ H_2O$
$\Delta H_c = 2\Delta H_{fCO_2} + 3\Delta H_{fH_2O(l)} - \Delta H_{fC_2H_6O_2}$
$\Delta H_c = 2\ (-94.1$ kcal/gmol) + 3 (-68.3 kcal/gmol) - (-109 kcal/gmol)
= **- 284.1 kcal/gmol**
ΔH_c = (- 284,100 cal/gmol) (1.8 Btu/lb/cal/g) (1/(62 g/gmol))
= **- 8200 Btu/lb GROSS Heating Value**

b. $C_2H_6O_2 + 5/2\ O_2 \rightarrow 2\ CO_2 + 3\ H_2O$
$\Delta H_c = 2\Delta H_{fCO_2} + 3\Delta H_{fH_2O(g)} - \Delta H_{fC_2H_6O_2}$
$\Delta H_c = 2\ (-94.1$ kcal/gmol) + 3 (-57.8 kcal/gmol) - (-109 kcal/gmol)
= **- 252.6 kcal/gmol**
ΔH_c = (- 252,600 cal/gmol) (1.8 Btu/lb/cal/g) (1/(62 g/gmol))
= **- 7330 Btu/lb NET Heating Value**

BC-5 Solution.

a. 5000 gal of waste with a [H$^+$] = X is diluted to 45,000 gal at pH = 7 or [H$^+$] = 10^{-7}. The minimum pH of 6 which can be tolerated is equivalent to a [H$^+$] = 10^{-6}; so from an ion balance:

BC-5 Solution (Continued).

$[H^+] = 10^{-6} = 5000\ (X/50,000) + (45,000) \times 10^{-7}$
$X = (50,000/5000) \times [10^{-6} - 45,000\ (10^{-7})/5000]$
$= \mathbf{0.91 \times 10^{-6}}$ **or pH = 5.04.**

b. With a tank volume of 100,000 gal, the solution is as follows:
$X = (100,000/5000) \times (10^{-6} - 95,000/100,000) \times 10^{-7}$
$= 1.81 \times 10^{-5}$, **pH = 4.74.**

BC-6 Solution.

a. The stoichiometric equation for the incineration of chlorobenzene is:

$$C_6H_5Cl + 7\ O_2 + 26.3\ N_2 \rightarrow 6\ CO_2 + HCl + 2\ H_2O + 26.3\ N_2$$

Therefore, 1 mol of chlorobenzene requires 7 mol of oxygen for complete combustion.

The chlorobenzene destruction rate is calculated as:

10^5 lb chlorobenzene/d x 1/1440 min/d = **69.44 lb chlorobenzene/min**
= 69.44 x 1/112.5 lb/lbmol = **0.617 lbmol chlorobenzene/min**

The oxygen requirement is:

$\dot{n}$ = 0.617 lbmol chlorobenzene/min x 7 lbmol O_2/lbmol chlorobenzene
= **4.32 lbmol O_2/min**

The air requirement is:

$\dot{n}$ = 0.617 lbmol chlorobenzene/min x (7 + 26.3) lbmol air/lbmol chlorobenzene
= **20.55 lbmol air/min**

The molecular weight of air is:

MW = 0.21 x 32 lb/lbmol O_2 + 0.78 x 28 lb/lbmol N_2 = **28.6 lb air/lbmol**

The air requirement at sea level then is:

Q_{sl} = 20.55 lbmol air/min (28.6 lb air/lbmol) = **586.9 lb air/min**
= 586.9 lb air/min/0.0756 lb air/ft^3 = **7763 cfm**

b. The air flow required in Albuquerque is calculated based on the ideal gas law with Albuquerque conditions of: elevation = 5200 ft, P = 12.1 psia, and T = 501.1 °R. Since

$$P_{sl}\ Q_{sl} = \dot{n}\ R\ T_{sl}$$

Then,

$$\frac{P_{sl}\ Q_{sl}}{T_{sl}} = \dot{n}R = \frac{P_{5200}\ Q_{5200}}{T_{5200}}$$

Solving for Q_{5200} yields:

$$Q_{5200} = \frac{14.7\text{ psia }(7763\text{ cfm})\ 501.1\ °R}{12.1\text{ psia }(520\ °R)} = \mathbf{9088\ cfm}$$

BC-6 Solution (Continued).

c. The particulate level allowed at 5200 ft elevation is based on the total mass flow of particulates:

0.08 gr/acf (7763 cfm) = [particulate] (9088 cfm)
[particulates] = 0.068 gr/acf

BC-7 Solution.

In order for the mercury to be removed by the filter, it must condense and form particles. Therefore, the question to be answered relates to the partial pressure of mercury compared to its vapor pressure at 150 °F.

Molar flow rate of Hg = (9.2 lb/h)/(200 lb Hg/lbmol) = **0.046 lbmol/h**

Mass flow rate of gas = 40,000 lb gas/(32 lb gas/lbmol) = **1250 lbmol/h**

y = lbmol Hg/(lbmol Hg + lbmol gas) = (0.046 lbmol/h)/(0.046 + 1250 lbmol/h)
= **3.68×10^{-5}**

Partial pressure = $y_i\ P_T$ = y (14.7 psia) = 3.68×10^{-5} (14.7 psia)
= **5.4×10^{-4} psia**

Since the partial pressure is much less than the vapor pressure, mercury will NOT condense and thus will NOT be removed by the filter.

BC-8 Solution.

a. 50 ppm is two standard deviations above the mean as calculated below:
(50 ppm - 20 ppm)/15 ppm = **2 std. above mean**

Using a normal distribution table, the probability of an event two standard deviations above the mean = **2.28%.**

b. Based on the assumptions presented, the receiving tank represents an average of over seven drums. The central limit theorem allows the expression of the standard deviation of the mean as the standard deviation (std) of the process divided by the square root of the sample number, i.e.,

std = 15 ppm/(7 drums)$^{0.5}$ = **5.7 ppm.**
= (25 ppm - 20 ppm)/5.7 ppm = **0.88 std above the mean**

Using a normal distribution table, there is an 18% chance of exceeding 25 ppm.

c. At a 95% confidence level, the value of the normal distribution is 0.05. Using a normal distribution table, this value represents **1.65 standard deviations above the mean.**

The number of equivalent drums, and thus the required minimum tank size is calculated from:

1.65 = (25 ppm - 20 ppm)/(15 ppm/$n^{0.5}$)
n = ((15 ppm/5 ppm) x 1.65)^2 = **24.5 drums**

Thus, the minimum tank size is: 24.5 drums x 50 gal/drum = **1,225 gal.**

BC-8 Solution (Continued).

d. The critical assumption made is that the Pb content of the drums and waste shipments are not correlated in time. In other words, it has been assumed that it is unlikely that the facility will receive consecutive shipments of waste with a Pb content higher than the mean. In actual practice, however, with multiple drums from a given source, it is highly likely that the drum content will be correlated. In addition, the distribution of Pb content in the problem was assumed to be normal; concentrations often have skewed distributions, e.g., lognormal.

Finally, it was assumed that the receiving tank is kept nearly full. Oversizing the tank would provide a margin of safety.

With additional information, e.g., autocorrelation and skewness of the concentration distribution, much more complex and realistic problems could be solved using queueing or inventory theory techniques.

CHAPTER TWO

Regulatory Concerns

R-1 Solution.

The permit process is designed to make maximum information available to the public through the public hearing process. It insures preparation and delivery of this information in an organized and descriptive manner. The permit process assures that all information used to choose a particular hazardous waste incineration system is public and available.

The permit process is designed to present to the regulators pertinent information which will allow them to make informed decisions in a reasonably consistent manner which, with proper coordination, utilizes a minimum of resources.

The application process requires basic information to demonstrate that the design and operation of the incinerator will result in the destruction of principle organic hazardous constituents (POHCs) by 99.99%; a hydrogen chloride scrubbing efficiency of 99% or < 4 lb/hr maximum HCl emissions; and, particulate emissions < 0.08 gr/dscf corrected to 50% EA.

The permit, in general, restricts the area of technical information that the permittee must develop, and most importantly, it serves as a shield of protection for the permittee against legal action so long as the permittee maintains compliance. This shield protects against arbitrary regulatory actions and unjustifiable public suits. It is important to note that it does not shield against actions which lead to endangerment of public health or the environment.

R-2 Solution.

Liability implies responsibility for an action. An individual may be held liable for a result if, in the mind of the normal, prudent person, the individual failed to exercise due caution. Examples include driving too fast, loosing control of a car, and causing damage to property or persons.

Strict liability implies responsibility without regard to prudence or care, i.e., without regard to negligence. Such standards are imposed for a variety of activities, such as handling dynamite, statutory rape, or hazardous waste management. These standards require that proper caution be exercised at all times. Defenses available, if harm results, are limited. These standards are the basis for training requirements imposed upon the permittee who is an owner/operator of hazardous waste treatment, storage or disposal facilities.

Joint and several liability is an assignment of responsibility when two or more persons fail to exercise the proper care and a division of harm is not possible. If two hunters fire their weapons and a person is killed, and there is no way to determine which projectile caused the harm, both hunters may be each held liable for the harm to the aggrieved party. In the case of joint and several strict liability, each party who managed a waste may be responsible for mitigating damages caused by the waste. For example, the generator, the transporter, the storage facility, and the incinerator operator may each individually or collectively be responsible for damages caused by mismanagement of a waste.

These provisions provide a tremendous impetus to hazardous waste generators to dispose of their waste on site under carefully controlled conditions. This concept of liability also burdens the generator with the threat of future costs as a result of someone else's improper actions. These values follow directly from our system of government which was created to assure that individual citizens do not suffer loss of property and freedoms (health) by the actions of others. The need for a careful choice of a contractor to carry out waste management and disposal responsibilities is also highlighted by these provisions.

R-3 Solution.

Two major reasons exist for this lack of enforcement/inspection of industrial waste volume/toxicity reduction program certification. The first was that the Congress was convinced that the EPA had neither enough personnel nor personnel with adequate expertise to evaluate industrial process systems. The second was that Congress was not convinced that this enforcement could be carried out without placing U.S. industry at a competitive disadvantage in world markets.

To resolve these problems, Congress did include an extensive research program in HSWA to augment the certification requirement, but left the responsibility of truthfulness with the industrial producers of hazardous waste. There does remain the basic requirement to truthfully report program activities by hazardous waste generators.

R-4 Solution.

Congress and EPA instituted a two phase permit system which allowed existing facilities (interim status permittees) to self enforce according to general requirements, while these facilities sought permits (Part B Permit applications) for permanent operation. These permits are issued for relatively long periods (five years) and are seldom modified.

Congress and EPA also minimized standards/permit modifications by requiring changes only when a treatment, storage or disposal facility is modified (i.e., process changes are being made) or expanded (new systems being added).

R-5 Solution.

The particulate concentration in the flue gas is

0.30 gr/8.0 acf = **0.0375 gr/acf**

Using Charles' law to calculate the volume at 60 °F = 520 °R:

8.0 acf x (520 °R/1040 °R) = **4.0 scf**

Particulate concentration = 0.30 gr/4.0 scf = **0.075 gr/scf**

The volume fraction of water in the flue gas is obtained from the mass composition of the waste and the balanced equation.

$C + H_2O + O_2 + N_2 \rightarrow CO_2 + H_2O + N_2$

For each 100 lb of waste (30 lb C and 70 lb water), the following number of lbmol result:

C: 30 lb/12 lb/lbmol = 2.5 lbmol
Water: 70 lb/18 lb/lbmol = 3.89 lbmol
Oxygen: 2.5 lbmol
Nitrogen: 79/21 (2.5 lbmol) = 9.40 lbmol

Fraction of water in flue gas: = lbmol water/total lbmol in flue gas
= lbmol H_2O/lbmol(CO_2 + H_2O + N_2) = 3.89 lbmol/(2.5 + 3.89 + 9.4 lbmol)
= **0.25**

Molar quantity of dry gas = (2.5 + 9.4 lbmol) = **11.9 lbmol**

Then the dry volume is calculated by subtracting the volume of water from the total gas volume: 4.0 scf (1-.25) = **3.0 dscf**

The particulate concentration on a dscf basis is then calculated as:
particulate concentrations = 0.30 gr/3.0 dscf = **0.10 gr/dscf**

R-5 Solution (Continued).

To correct to 50% EA, 50% more N_2 and O_2 are added to the flue gas:

Excess nitrogen added = 0.5 (79/21) (2.5 lbmol) = **4.7 lbmol**
Excess oxygen added = 0.5 (2.5) = **1.25 lbmol**

Then the total dry flue gas molar quantity is increased by the EA volume to yield a total quantity at 50% EA of:

Total flue gas molar quantity = 2.5 + 9.4 + 4.7 + 1.25 lbmol
= **17.85 lbmol**

Total flue gas volume can the be calculated based on the ratio of total moles of flue gas at 50% and 0% EA:

Flue gas volume at 50% EA = 3.0 dscf (17.85 lbmol/11.89 lbmol)
= **4.5 dscf corrected to 50% EA**

The particulate concentration is then calculated on a dry weight basis, corrected for 50% EA as:

Particulate concentration = 0.30 gr/4.5 dscf corrected to 50% EA
= **0.067 gr/dscf corrected to 50% EA**

Since this does not exceed the particulate standard, **the incinerator is in compliance.**

R-6 Solution.

The particulate concentration in the stack is:

Particulate concentration = 0.16 g collected/35 dscf sampled (15.43 gr/g)
= **0.07 gr/dscf**

Since this does exceed the particulate standard of 0.05 gr/dscf, **the incinerator is not in compliance.**

The actual particulate emission rate is the product of the stack flow rate and the stack flue gas particulate concentration. The stack flow rate is calculated from the velocity measurements provided in the problem statement using the second velocity equation given:

$$v = 0.85 \sqrt{2\ (9.81\ m/s^2) \left(\frac{1000\ kg/m^3}{1.084\ kg/m^3}\right) 0.0254\ h}$$

$$= 0.85\ (21.437)\ \sqrt{h_{ave}}$$

$$= 0.85\ (21.437)\ (.6142)$$

$$= \mathbf{11.2\ m/s = 36.75\ fps}$$

Stack flow rate = v (cross sectional area) = 36.75 fps (π/4) (2 ft)^2
= 115.45 acfs = **6924 acfm**

Dry volumetric flow rate = (1 - 0.07) x 6924 acfm = **6439 dacfm**

Correct to standard conditions 70 °F and 1 atm pressure:

R-6 Solution (Continued).

Standard volumetric flow rate = 6439 dcfm (530 °R/600 °R) 29.6 psi/29.9 psi
= 5631 dscfm

Particulate emission rate = 0.07 gr/dscf x 5631 dscfm = **394 gr/min**
= 394 gr/min x 1 lb/7000 gr = **0.056 lb/min**
= 0.056 lb/min x 1440 min/d = **81.1 lb/d**

Molecular weight of flue gas is based on the mole fraction of the flue gas components. The flue gas is 7% water, and 93% other components by volume. On a dry gas basis, the flue gas molecular weight is:

Molecular weight = 0.07 O_2 (32 lb/lbmol) + 0.14 CO_2 (44 lb/lbmol) + 0.79 N_2 (28 lb/lbmol) = **30.52 lb/lbmol**

The average molecular weight of the stack gas on an actual (wet) basis is then:

Average MW = 0.07 water (18 lb/lbmol) + 0.93 other components x 30.52 lb/lbmol
= **29.64 lb/lbmol**

R-7 Solution.

The gas volume must be corrected for water content, temperature and oxygen content since the flue gas does not correspond to the reference conditions.

Correction for water content:

$$V_2 = V_1\left(\frac{n_2}{n_1}\right) = 1.0\left(\frac{n_1 - n_{H_2O}}{n_1}\right) = \frac{1.0 - 0.25}{1.0} = \mathbf{0.75\ dacf}$$

Correction for oxygen content is based on 50% excess air provided above the stoichiometric level required for complete combustion. The stoichiometric oxygen required is based on carbon dioxide formation using the following balanced equation:

$$C + 1.5\ O_2 \rightarrow CO_2 + 0.5\ O_2$$

Thus, the flue gas should contain 0.5 volumes of oxygen per volume of carbon dioxide at 50% EA. The actual amount is 1 volume oxygen per volume of carbon dioxide, and thus the actual volume must be reduced using the ideal gas law:

$$V_3 = V_2\left(\frac{n_3}{n_2}\right) = 0.75\left(\frac{n_2 - n_{O_2}}{100}\right) = 0.75\left(\frac{100 - 6.25}{100}\right)$$

= 0.70 dacf corrected to 50% EA

Temperature corrections are made based on Charles' Law:

$$V_4 = V_3\left(\frac{T_4}{T_3}\right) = 0.70\left(\frac{460 + 68}{460 + 450}\right) = \mathbf{0.41\ dscf\ corrected\ to\ 50\%\ EA}$$

R-7 Solution (Continued).

Loading under dscf corrected to 50% EA conditions is determined as follows:

$$L_{dscf} = L_{acf}\left(\frac{V_{acf}}{V_{dscf}}\right) = 0.06 \text{ gr/acf}\left(\frac{1.0 \text{ acf}}{0.41 \text{ dscf}}\right)$$

$$= \mathbf{0.15 \text{ gr/dscf corrected to 50\% EA}}$$

This value exceeds the regulatory maximum limit of 0.08 gr/dscf corrected to 50% excess air, and the **unit is therefore out of compliance.**

R-8 Solution.

a. Use Dulong's equation to calculate the NHV of the mixture in Btu/lb. Dichlorobenzene ($C_6H_4Cl_2$) has the following composition of elemental constituents:

Carbon: 6 (12 lb/lbmol) = 72 lb/lbmol
Hydrogen: 4 (1 lb/lbmol) = 4 lb/lbmol
Chlorine: 2 (35.45 lb/lbmol) = 70.9 lb/lbmol
Total Molecular Weight: 72 + 4 + 70.9 = 146.9 lb/lbmol

$$NHV = 14{,}000\ m_c + 45{,}000\ (m_H - 1/8\ m_o) - 760\ m_{Cl} + 4500\ m_s$$

where		dichlorobenzene	Coal	Oil	
	mc	= (72/146.9) 0.4	+ 0.793 (0.4)	+ 0.688 (0.2)	= 0.651
	mH	= (4/146.9) 0.4	+ 0.038 (0.4)	+ 0.078 (0.2)	= 0.042
	mo	=	+ 0.053 (0.4)	+ 0.020 (0.2)	= 0.025
	mCl	= (70.9/146.9) 0.4	+	+ 0.10 (0.2)	= 0.213
	ms	=	+ 0.011 (0.4)	+ 0.015 (0.2)	= 0.0074

NHV = 14,000 Btu/lb (0.651) + 45,000 Btu/lb (0.042 - 1/8 (0.025)) - 760 Btu/lb (0.213) + 4500 Btu/lb (0.0074) = **10,735 Btu/lb**

b. The maximum feed rate is

$$\dot{w} = \frac{25{,}000 \text{ Btu/h-ft}^3 \left(\pi \dfrac{5^2}{4}\right) \text{ft}^2 \ 15 \text{ ft}}{(1-0.08\ \{\text{heat loss}\})\ 10{,}735 \text{ Btu/lb}} = 745.5 \text{ lb/h}$$

Weighted average specific gravity of the mixed waste stream:

Mean specific gravity = 0.4 (1.53) + 0.4 (2.3) + 0.2 (0.75) = 1.682

Density of mixed waste stream = 63.4 lb/ft^3 (1.682) = 105 lb/ft^3

Waste stream flow rate = 745.5 lb/h (1 ft^3/105 lb) (7.48 gal/ft^3) (1 h/60 min)

Q = 0.885 gpm

R-8 Solution (Continued).

c. The incinerator exhaust temperature is given as:

$$T = 60 + \frac{10{,}735\ (0.92)}{0.3\ [1 + 1.5\ (7.54 \times 10^{-4})\ (10{,}735)\ (0.92)]}$$

$$T = \mathbf{2705\ ^\circ F}\ ;\ T = 273 + 5/9\ (2705 - 32) = \mathbf{1758\ K}$$

d. The HCl equilibrium constant at the incinerator exhaust temperature is given as:

$$K_p = 0.23 \times 10^{-3} \exp(7340/1758) = \mathbf{0.0149\ atm^{-1/2}}$$

The distribution of Cl_2 in the flue gas is calculated based on the exhaust gas composition on a molar basis. The molar flow rate of each component from the HWI is:

C: 0.651 lb C/lb mixed waste (745.5 lb mixed waste/h) (1 lbmol C/12 lb) = **40.44 lbmol/h**
H_2: 0.042 lb H/lb mixed waste (745.5 lb mixed waste/h) (1 lbmol H_2/2 lb) = **15.66 lbmol/h**
O_2: 0.025 lb O/lb mixed waste (745.5 lb mixed waste/h) (1 lbmol O_2/32 lb) = **0.582 lbmol/h**
Cl: 0.213 lb Cl/lb mixed waste (745.5 lb mixed waste/h) (1 lbmol O_2/71 lb) = **2.24 lbmol/h**
S: 0.0074 lb S/lb mixed waste (745.5 lb mixed waste/h) (1 lbmol S/32 lb) = **0.172 lbmol/h**
N_2: [0.01 (0.4) coal + 0.017 (0.2) oil] lb N_2/lb mixed waste (745.5 lb mixed waste/h) (1 lbmol N_2/28 lb) = **0.197 lbmol/h**

For stoichiometric combustion conditions

$$40.44\ C + 15.66\ H_2 + 0.582\ O_2 + 0.172\ S + 0.197\ N_2 + 2.24\ Cl_2 + a(O_2 + 3.76\ N_2) \rightarrow 40.15\ CO_2 + e\ H_2O + f\ HCl + g\ SO_2 + 3.76\ a\ N_2$$

where

H: 2 (15.66) = 2 e + f; e = 13.42

O: 2 (0.582) + 2 a = 40.44 (2) + e + 2g; a = 46.74

Cl: 2 (2.24) = f = 4.48

S: 0.172 = g

For 50% EA a = 1.5 (46.74) = 70.11

For actual combustion conditions

$$40.44\ C + 15.66\ H_2 + 0.172\ S + 2.24\ Cl_2 + (0.582 + 70.11)\ O_2 + [0.197 + 70.11\ (3.76)]\ N_2 \rightarrow 40.44\ CO_2 + p\ H_2O + 4.48\ HCl + 70.11\ (3.76)\ N_2 + q\ O_2 + 0.172\ SO_2$$

H: 15.66 (2) = 2 p + 4.48; p = 13.42

O: 2 (70.692) = 2(40.44) + p + 2q + 2(0.172); q = 23.54

R-8 Solution (Continued).

HWI exhaust gas composition compilation

Component	lbmol/h	Mole Fraction
CO_2	40.44	0.1170
H_2O	13.42	0.039
HCl	4.48	0.0130
N_2	263.6	0.7626
O_2	23.54	0.0681
SO_2	0.172	0.0005
	346.8	

Let Z be partial pressure of Cl_2 at equilibrium

$p_{H_2O} = 0.039 + Z$, $p_{HCl} = 0.0130 - 2\ Z$, $p_{O_2} = 0.0681 - 0.5\ Z$

$$0.0149 = \frac{(0.039 + Z)\ Z}{(0.013 - 2\ Z)^2\ (0.0681 - 0.5\ Z)^{0.5}}$$

Solving for Z yields:

$Z = 16.8 \times 10^{-6}$, so that $[Cl_2]$ = **16.8 ppm**

e. The removal efficiency to meet the HCl standard is calculated based on the molar HCl emission rate from the HWI. This is calculated from the exhaust gas composition table presented above.

HCl mass emission rate = 4.48 lbmol/h (36.45 lb/lbmol) = 163.3 lb/h

As the quencher removes 90% of the HCl in the incinerator exhaust gas, the mass of HCl loaded to the scrubber is:

HCl mass to scrubber = 163.3 lb/h (1 - 0.9) = **16.3 lb/h**

The required removal efficiency is then:

Removal efficiency = 100 (1 - 4/16.3) = **75.5%**

f. The residence time is calculated based on the molar emission rate of the HWI exhaust gas and using the ideal gas law:

$$Q = \dot{V} = \frac{\dot{n}\ R\ T}{P} = \frac{345.45\ \text{lbmol/h}\ (10.73\ \text{psi-ft}^3/\text{lbmol-°R})\ (2705+460\ °\text{R})}{14.7\ \text{psi}}$$

$$Q = 798{,}071\ \text{acfh} = 13{,}301\ \text{acfm}$$

Retention time = V/Q = $(\pi\ 5^2/4)\ 15\ \text{ft}^3/13{,}301$ acfm = 0.022 min = **1.33 s**

g. Critique:

Technology: need scrubber for particulate removal, need to increase residence time with larger incinerator or reduced feed rate
Emissions: SO_2 emissions should be checked for compliance
Energy: 2705 °F; heat recovery may be of interest
Economy: Consider venturi scrubber for particulate and chlorine gas removal instead of packed bed scrubber

R-9 Solution.

a. The DRE for each hazardous material is related to its mass flow rate into (i) the HWI and the mass flow rate out (o) of the stack by the equation:

$$DRE = \left[1 - \frac{m_o}{m_i}\right] 100$$

where m_o is equal to the product of the mass concentration in the stack, ρ_o, and the volumetric flow rate, Q_o, out the stack. Each stack gas concentration, C_o, in ppm can be shown to be related to ρ_o by the following equation at 25°C and 1 atm pressure:

$$\rho_o = 40.9\ C_o\ MW$$

where ρ_o is in $\mu g/m^3$ and MW signifies the gram molecular weight.

The influent mass of each compound is 5% of the influent waste feed = 0.05 (5000 lb/h) = **250 lb/h**

With the above information the table below was constructed.

	m_i lb/h	MW	C_o ppm	ρ_o g/m^3	m_o lb/min	DRE %
C_2HCl_3	250	131.5	0.03	161.35	2.147×10^4	99.9999
C_2Cl_4	250	166	0.02	135.79	1.807×10^4	99.9999
$CHCl_2F$	250	103	0.30	1263.8	1.682×10^{-3}	99.9993
$C_8H_4Cl_2O$	250	187	1.8	13766.9	1.83×10^{-2}	99.9927

b. Assume F and Cl have essentially the same influence on the waste NHV, then the mass fraction of each element in the waste compounds can be calculated as:

Mass flow rate of C from the C_2HCl_3 is: 2 (12 lb/lbmol)/131.5 lb/lbmol (250 lb/h) = **45.6 lb/h**

The remaining values in the table below are calculated in an identical manner.

	m_i lb/h	MW	C	H	Cl & F	O
C_2HCl_3	250	131.5	45.6	1.9	202.5	
C_2Cl_4	250	166	36.1		213.9	
$CHCl_2F$	250	103	29.13	2.43	218.45	
$C_8H_4Cl_2O$	250	187	128.3	5.35	94.92	21.4
			239.1	9.68	729.77	21.4

In this table, a basis of 1000 lb/h total waste flow of the four compounds was used. The NHV can be calculated for this four compound mixture using Dulong's equation:

$$NHV = 14{,}000\ (0.239) + 45{,}000\ [0.0997 - 1/8\ (0.0214)] - 760\ (0.7298)$$
$$= \mathbf{3107\ Btu/lb}$$

R-9 Solution (Continued).

With the mixture of the four compounds and the #2 fuel oil, the NHV of the entire waste mixture can be calculated as follows:

$NHV_{mixture}$ = 0.8 (18,650 Btu/lb fuel oil) + 0.2 (3107 Btu/lb waste compounds) = **15,541 Btu/lb**

R-10 Solution.

a. The incinerator receives 5 t/h of waste with 2% Cl content. Assuming all chlorine is converted to HCl, the amount of HCl formed is:

Cl in feed = 5 t waste/h x 2000 lb/t x 0.02 lb Cl/lb waste = **200 lb Cl/h**
HCl formed = 200 lb Cl/h x 36.5 lb HCl/35.5 lb Cl = **205.6 lb HCl/h**

The maximum permissible mass emission rate of HCl at 99% control is calculated as:

= 205.6 lb HCl/h (1 - 0.99) = **2.06 lb HCl/h emitted**

b. To calculate particulate emissions at 50% EA, a 1 mole C basis is used along with the following balanced equation:

$$C + 1.5\ O_2 + 1.5\ (79/21)\ N_2 \rightarrow CO_2 + 0.5\ O_2 + 1.5\ (79/21)\ N_2$$

The volume of flue gas generated per lbmol of C combusted is calculated from the ideal gas law:

$$V = \frac{n\ R\ T}{P} = \frac{(1+0.5+1.5(3.74)\ \text{lbmol})(460+68\ ^\circ\text{R})(0.7302\ \text{ft}^3\text{-atm/lbmol-}^\circ\text{R})}{1\ \text{atm}}$$

= **2741 dscf/lbmol C**

Based on the required incinerator emission standard of 0.07 gr/dscf, the maximum emission rate allowed at 50% EA is as follows:

Volume of gas generated per hour:
= 5 t fuel/h (0.8 lb C/lb fuel) (2000 lb/t) (2741 dscf/12 lb C)
= **1,827,333 dscfh**

Mass of particulates allowed to discharge per hour:
= 1,827,333 dscfh (0.07 gr particulates/dscf) (1 lb/7000 gr)
= **18.27 lb particulates/h**

The actual particulate emission rate is:

Mass emitted = 5 t fuel/h (2000 lb/t) (0.05 lb ash/lb fuel)
= **500 lb particulates/h, well above the allowable limit, requiring particulate removal to meet permit requirements.**

c. The actual combustion efficiency can be calculated based on the ratio of carbon dioxide to total effluent carbon species in the flue gas effluent, and is:

$$\text{Combustion Efficiency} = \frac{12}{12 + 0.0020}\ (100) = 99.98\%$$

R-11 Solution.

a. DREs for each compound are calculated as follows:

Toluene:
100 (860 lb in/h -0.2 lb out/h)/(860 lb in/h) = **99.98%**

Chlorobenzene:
100 (450 lb in/h - 0.02 lb out/h)/(450 lb in/h) = **99.996%**

Dichlorobenzene:
100 (300 lb in/h - 0.03 lb out/h)/(300 lb in/h) = **99.99%**

Molar HCl production is calculated based on all Cl coming into the incinerator, and assuming that all Cl is converted to HCl upon combustion

For chlorobenzene, the amount of HCl produced is calculated assuming:

$$C_6H_5Cl + 7\ O_2 \rightarrow 6\ CO_2 + HCl + 2\ H_2O$$

HCl produced from chlorobenzene
= 450 lb/h/(6 (12 lb C/lbmol) + 5 (1 lb H/lbmol) + 1 (35.45 lb C/lbmol))
= **4 lbmol HCl/h**

For dichlorobenzene, the amount of HCl produced is calculated assuming:

$$C_6H_4Cl_2 + 6.5\ O_2 \rightarrow 6\ CO_2 + 2\ HCl + 2\ H_2O$$

HCl produced from dichlorobenzene
= 300 lb/h/(6 (12 lb C/lbmol) + 4 (1 lb H/lbmol) + 2 (35.45 lb C/lbmol))
= **4.08 lbmol HCl/h**

Therefore, the total amount of HCl produced
= (4 + 4.08 lbmol) (36.45 lb/lbmol) = **294.5 lb HCl/h**

The DRE for HCl is:
100 (294.5 lb HCl in/h - 4.2 lb HCl out/h)/294.5 lb HCl in/h = **98.57%**

b. The amount of Pb in the waste mixture is based on the TOC content of the waste as given by:

TOC for toluene (C_7H_8):

$$\%\ \text{TOC in toluene} = \frac{\text{Mass of C in toluene}}{\text{Total mass of toluene}} = \frac{7\ (12)}{7\ (12) + 8\ (1)}$$

$$= 0.913$$

Toluene TOC emission rate = (0.913) 860 lb toluene/h = **785 lb TOC/h**

TOC for chlorobenzene (C_6H_5Cl):

$$\%\ \text{TOC in chlorobenzene} = \frac{\text{Mass of C in CB}}{\text{Total Mass of CB}} = \frac{6\ (12)}{6\ (12) + 5\ (1) + 1\ (35.45)}$$

$$= 0.64$$

R-11 Solution (Continued).

Chlorobenzene TOC emission rate = (0.64) 450 lb CB/h = **288 lb TOC/h**

TOC for dichlorobenzene ($C_6H_4Cl_2$):

$$\% \text{ TOC in dichlorobenzene} = \frac{\text{Mass of C in DCB}}{\text{Total Mass of DCB}} = \frac{6\ (12)}{6\ (12) + 4\ (1) + 2\ (35.45)}$$

$$= 0.49$$

Chlorobenzene TOC emission rate = (0.49) 300 lb DCB/h = **147 lb TOC/h**

The amount of Pb in the flue gas is:

Total emission of TOC = (785 + 288 + 147 lb TOC/h) = **1220 lb TOC/h**
Total mass of Pb = 1220 lb TOC/h (0.005 lb Pb/lb TOC) = **6.1 lb Pb/h**

Concentration of Pb in the incinerator flue gas is
= 6.1 lb Pb/h (1 h/60 min)/14,280 dscfm = **7.12 x 10^{-6} lb/dscf**
= 7.12 x 10^{-6} lb/dscf (1.6033 x 10^{10} μg/m^3/lb/dscf) = **114,147 μg/m^3**

The particulate concentration in the flue gas is calculated as follows:

Total emission of particulates = 9.65 lb/h (7000 gr/lb) = **67,550 gr/h**

Concentration of particulates in the incinerator flue gas is
= 67,550 gr/h/(14,280 dscfm x 60 min/h) = **0.079 gr/dscf**

c. The unit **is in compliance with present Federal regulations for POHCs and particulates,** but is **NOT in compliance for HCl emissions.**

CHAPTER THREE

Stoichiometry

S-1 Solution.

The general balanced equation for the complete combustion of the alkyl-dichlorobenzenes is as follows:

$$C_nH_{2n-8}Cl_2 + (1.5n - 2.5)\ O_2 \rightarrow n\ CO_2 + 2\ HCl + (n - 5)\ H_2O, \text{ where } n > 6$$

S-2 Solution.

a. Per lb of waste:

Cl:molar flow rate of Cl = 0.035 lb/(35.45 lb/lbmol Cl) = 0.001 lbmol Cl. With a stoichiometry according to the following equation:

$$2\ Cl + H_2 \rightarrow 2\ HCl$$

the following calculation for the amount of hydrogen used in this reaction results:

$$-0.001 \text{ lbmol Cl}/2 = -0.0005 \text{ lbmol } H_2$$

H: molar flow rate of H = 0.109 lb/(1.00 lb/lbmol H) = 0.109 lbmol H. With a stoichiometry according to the following equation:

$$2\ H_2 + O_2 \rightarrow 2\ H_2O$$

the following calculation for the amount of water produced in this reaction results:

$$0.109 \text{ lbmol H}/2 = 0.0545 \text{ lbmol } H_2O \text{ produced}$$

Since one water is produced from each H_2, the total amount of water produced = 0.0545 - 0.0005 = 0.0540 lbmol, and the lb of water produced is 0.54 lbmol (18 lb/lbmol) = 0.972 lb water.

Then the total water in the flue gas is 0.972 lb + 0.022 lb **= 0.994 lb water/lb waste in the flue gas before the scrubber**

b. Per lb of waste:

H: molar flow rate of H = 0.109/2 = 0.054 lbmol H_2. With a stoichiometry according to the equation above, 0.5 moles of oxygen are required per mole of H_2, or 0.054/2 = 0.027 lbmol O_2 are required for the hydrogen in the waste.

C: molar flow rate of C = 0.732/12 = 0.061 lbmol C. With a stoichiometry according to the equation below:

$$C + O_2 \rightarrow 2\ CO_2$$

1 mole of oxygen is required per mole of C, or 0.061 lbmol O_2 are required for the carbon in the waste.

The total amount of oxygen is calculated as:

$$0.027 \text{ lbmol} + 0.061 \text{ lbmol} - 0.102 \text{ lb } O_2/(32 \text{ lb/lbmol } O_2) = 0.0849 \text{ lbmol}$$

The volume of oxygen is 0.0848 lbmol (379.5 scf/lbmol) = **32.2 scf/lb waste**

S-2 Solution (Continued).

c. 0.001 lbmol Cl were generated from the waste to produce 0.001 lbmol HCl as seen in Part a above. Based on the stoichiometry of soda ash neutralization:

$$Na_2CO_3 + 2\ HCl \rightarrow 2\ NaCl + H_2CO_3$$

2 moles of HCl are neutralized by one mole of soda ash, therefore, **0.001/2 = 0.0005 lbmol of soda ash/lb waste are required for neutralization.**

S-3 Solution.

The balanced stoichiometric reaction for dichlorobenzene is shown below:

$$C_6H_4Cl_2 + 6.5\ O_2 \rightarrow 6\ CO_2 + H_2O + 2\ HCl$$

Therefore, for 1 lb of dichlorobenzene, the following mass of HCl is produced:

[1 lb/(147 lb/lbmol DCB)] [(2 mol HCl/mol DCB)] = **0.0136 lbmol HCl produced**

The balanced stoichiometric reaction for tetrachlorobiphenyl is as shown below:

$$C_{12}H_6Cl_4 + 12.5\ O_2 \rightarrow 12\ CO_2 + H_2O + 4\ HCl$$

Therefore, for 1 lb of tetrachlorobiphenyl, the following mass of HCl is produced:

[1 lb/(290 lb/lbmol TCB)] [(4 mol HCl/mol DCB)] = **0.0138 lbmol HCl produced**

Thus, the amount of acid produced does not change significantly (≈ 1.5%), and neither will the amount of base required for neutralization.

S-4 Solution.

a. Assume that 50 mol of Cl_2 and 50 mol of H_2O enter the reactor during a unit period of time. Let Z = mol HCl removed in the scrubber during this time; X = mol H_2O + mol Cl_2 in the recycle stream before release; and Y = mol O_2 in the recycle stream before release.

An overall Cl_2 balance can be written as:

$$50\text{ mol (in)} = Z/2\text{ mol (out as HCl)} + 0.01\ X/2\text{ (out as release gas)}$$

An overall O_2 balance can be written as:

$$50\text{ mol (in as } H_2O) = 0.01\ X/2\text{ (out as } H_2O) + 0.01\ Y\text{ (out as } O_2)$$

Because the Cl_2 molar composition of the release gas is 0.1, the following expression can be written:

$$\frac{\frac{X}{2}}{X + Y} = 0.1$$

S-4 Solution (Continued).

From these three equations, X, Y and Z can be found, yielding values of:

X = 588 mol water + mol chlorine in recycle stream, Y = 2353 mol oxygen in recycle stream, and Z = 94.2 mole HCl removed in scrubber

Then, the recycle ratio = 0.99 (X + Y)/100 = **29.1**

b. Molar composition of the reactor products are calculated according to the following expressions:

$$\text{mole fraction of HCl} = \frac{Z}{X + Y + Z} = 0.031$$

$$\text{mole fraction of } O_2 = \frac{Y}{X + Y + Z} = 0.775$$

$$\text{mole fraction of } H_2O = \frac{X/2}{X + Y + Z} = 0.0969$$

$$\text{mole fraction of } Cl_2 = \frac{X/2}{X + Y + Z} = 0.0969$$

c. The equilibrium constant for the HCl/Cl_2 reaction presented in the problem statement can be written as:

$$K_p = \frac{p(HCl)^2 \; p(O_2)^{1/2}}{p(Cl_2) \; p(H_2O)} = \frac{mol(HCl)^2 \; mol(O_2)^{1/2}}{mol(Cl_2) \; mol(H_2O)}$$

$$= \frac{(\text{mol fraction HCl})^2 \; (\text{mol fraction } O_2)^{1/2}}{(\text{mol fraction } Cl_2) \; (\text{mol fraction } H_2O)}$$

because

$$\frac{n_{total}^{1/2} \; R^{1/2} \; T^{1/2}}{V^{1/2}} = P_{total}^{1/2} = 1 \text{ atm}$$

Thus,

$$\ln(K_p) = \ln \frac{Z^2 \; Y^{1/2}}{(X/2)^2} = 1.61$$

However, $\ln(K_p)$ is given by the following expression:

$$-\ln(K_p) = \frac{7048.7}{T} + 0.151 \ln(T) - 9.06 \times 10^{-5} \; T - 27{,}141 \; T^{-2} - 8.09$$

Substituting for $\ln(K_p)$ and solving for T by trial and error yields a calculated reactor temperature of **T = 1274 K = 1001 °C.**

S-5 Solution.

A material balance is written showing O_2 demand (air) and flue gas products on the basis of 100 lb of sludge (wet) combusted.

component	lbmol of atoms	reaction	gaseous products (lbmol)
40 lb H_2O	2.22		2.22 (as H_2O)
30 lb C	2.5	$C + O2 \rightarrow CO2$	2.5 (as CO_2)
6 lb H	3	$H_2 + 0.5\ O_2 \rightarrow H_2O$	3 (as H_2O)
2 lb Cl	0.029	$Cl_2 + H_2O \rightarrow 2\ HCl + 0.5\ O_2$	0.029 (as Cl_2)
8 lb O	0.25		0.25 (as O_2)
4 lb N	0.14		0.14 (as N_2)
10 lb inert			
Y lbmol CH_4		$CH_4 + 2\ O_2 + 7.52\ N_2 \rightarrow CO_2 + 2\ H_2O + 7.52\ N_2$	

Then the total O_2 demand in lbmol is

2.5 (for CO_2) + 0.5 (3) (for H_2O)- 0.5 (0.029) (from Cl_2) - 0.25 (present in sludge) = 3.74 lbmol

The corresponding amount of N_2 from the air is 14.07 lbmol, which appears in the exhaust gas along with the gaseous products indicated above.

Next the heat production is considered. All heat released (q_1) goes to raise the temperature of the exhaust gas (q_2). Thus, $q_1 = q_2$.

For X ft^3 of CH_4 per 100 lb of sludge at 60 °F:

q_1 = (1000 Btu/lb waste) (100 lb waste) + (X ft^3 CH_4)(1000 Btu/ft^3 CH_4)
= 100,000 Btu + Y lbmol CH_4 (379 ft^3 CH_4/lbmol CH_4) (1000 Btu/ft^3 CH_4)

On the basis of 100 lb of wet sludge, we find q_2 using the lbmol of each component and the temperature-averaged heat capacities (from Appendices C and F) for each component of the exhaust gas including the inert particulate material:

q_2 = lbmol (H_2O) [2000 °F (9.33 Btu/lbmol-°F) - 60 °F (8.0 Btu/lbmol-°F)]
+ lbmol (CO_2) [2000 °F (11.94 Btu/lbmol-°F) - 60 °F (8.6 Btu/lbmol-°F)]
+ lbmol (HCl) [2000 °F (7.38 Btu/lbmol-°F) - 60 °F (6.93 Btu/lbmol-°F)]
+ lbmol (N_2) [2000 °F (7.53 Btu/lbmol-°F) - 60 °F (6.94 Btu/lbmol-°F)]
+ lb (inerts) (1.5 Btu/lb)(2000 °F - 60 °F)

q_2 = 18,180 Btu/lbmol H_2O [2.22 lbmol H_2O + 3.0 lbmol (H) - 0.029 lbmol (Cl)
+ 2 Y lbmol (CH_4)] + 23,364 Btu/lbmol CO_2 [2.5 lbmol (C) + Y lbmol (CH_4)]
+ 14,344 Btu/lbmol HCl [2 (0.029 lbmol) (Cl)]
+ 14,644 Btu/lbmol N_2 [0.14 lbmol (N_2) + 14.07 lbmol (air)
+ 7.52 Y lbmol (CH_4)] + 29,100 Btu/lb inerts (10 lb inerts)

Equating q_1 and q_2 and solving for Y yields Y = 2.64 lbmol CH_4 required per 100 lb of sludge.

S-5 Solution (Continued).

The incinerator heat load using the calculated q_1 is found to be:

q_1 = 1.1 x 10^6 Btu = (1.1 x 10^6 Btu) (2000 lb/t) (3 t/h) = 66 x 10^6 Btu/h

Note that the incinerator is properly sized to handle this load at a rating of 75 x 10^6 Btu/h.

Finally, the amount of natural gas required is:

66 x 10^6 Btu/h - (3 t/h) (1000 Btu/lb) (2000 lb/t) = 60 x 10^6 Btu/h or

60 x 10^6 Btu/h/1000 Btu/ft^3 CH_4 = **1000 acfm at 60 °F**

S-6 Solution.

a. The concentration of water in the combustion air is calculated using data from Appendix A:

[H_2O] = 0.0102 lb H_2O/lb dry air at 70 °F and 65% RH
= (0.0102 lb H_2O/lb dry air) (28.8 lb/lbmol air/18 lb/lbmol H_2O)
= **0.0163 lbmol H_2O/lbmol dry air**

From this value, the amount of water in the flue gas that resulted from the water in the combustion air can be calculated:

The lbmol of H_2O in each (lbmol of O_2 + 3.76 lbmol of N_2) is 0.0163 lbmol H_2O/lbmol dry air

lbmol H_2O/lbmol dry air = (0.0163 lbmol H_2O /lbmol dry air) (1 + 3.763 lbmol dry air) = **0.0776 lbmol H_2O /lbmol dry air**

The balanced, stoichiometric equations for these compounds are shown as follows:

C_6H_5Cl + 7 (O_2 + 3.76 N_2) + 0.0978 H_2O → 6 CO_2 + 2 H_2O + HCl + 22.56 N_2

CH_2Cl_2 + O_2 + 3.76 N_2 + 0.0163 H_2O → CO_2 + 2 HCl + 3.76 N_2 + 0.0163 H_2O

Thus, at 100% excess air, the two balanced combustion equations are:

C_6H_5Cl + 14 O_2 + 52.6 N_2 + 1.09 H_2O → 6 CO_2 + 3.09 H_2O + HCl + 52.6 N_2 + 7 O_2

CH_2Cl_2 + 2 O_2 + 7.51 N_2 + 0.155 H_2O → CO_2 + 2 HCl + O_2 + 7.51 N_2 + 0.155 H_2O

At 90 mol% C_6H_5Cl, 10 mol% CH_2Cl_2 and 100% excess air, the combined balanced equation is:

0.9 C_6H_5Cl + 0.1 CH_2Cl_2 + [0.9 (14.0) + 0.1 (2.0)] O_2+ [0.9 (52.6) + 0.(7.51)] N_2 + [0.9 (1.09) + 0.1 (0.155)] H_2O →
[0.9 (6.0) + 0.1 (1.0)] CO_2 + [0.9 (1.0) + 0.1 (2.0)] HCl + [0.9 (7.0) + 0.1 (1.0)] O_2 + [0.9 (52.6) + 0.1 (7.51)] N_2 + [0.9 (3.09) + 0.1 (0.155)] H_2O

Then, the mole fraction of HCl in the flue gas is:

0.9 + 0.2 mol HCl/(total moles of products) = 1.1 mol/(63.884 mol) = **0.0172**

S-6 Solution (Continued).

b. The molecular weight of the flue gas is:

$$MW = \frac{\sum_{i=1}^{n} n_i (MW)_i}{\sum_{i=1}^{n} n_i}$$

= [5.5 (44) + 1.1 (63.45) + 6.4 (32) + 28 (48.09) + 2.794 (18)]/63.884
= **29.49 lb/lbmol of flue gas**

For 0.9 mol of C_6H_5Cl and 0.1 mole of CH_2Cl_2, there are 109.7 lb of chlorinated organics combusted and 5.48 lb of particulate material entrained in the flue gas. The total volume of flue gas produced at 100% excess air is obtained from the ideal gas law:

V=nRT/P = 63.884 lbmol (0.7302 atm ft^3/lbmol-°R) (460 + 68 °R)/1 atm
= **24630 scf (wet)**

The volume of dry gas is:

mol H_2O/total mol of flue gas = [0.9 (3.09) + 0.1 (0.155)]/63.884 = **0.044**

V(dry) = 24,630 ft^3 (1 -0.044) = **23,600 dscf**

The mole ratio for combustion air at 50% excess air to that at 100% excess air is:

$$\frac{1.1 + 5.5 + 6.4/2 + 48.09/2 + 2.794}{1.1 + 5.5 + 6.4 + 48.09 + 2.794} = 0.5735$$

Then, V (dry, 50% excess air) = 23,600 ft^3 (0.5735) = 13,500 ft^3

So the mass concentration of particulate material at standard conditions, 50% excess air, and with dry air is:

total mass of particles/V (dry, 50% excess air) = 5.48 lb (7000 gr/lb)/13,500 ft^3 = **2.83 gr/dscf corrected to 50% excess air**

Finally, the required particulate mass collection efficiency is:

E = [(2.83 - 0.08)/2.83] x 100 = **97.2%**

S-7 Solution.

It can be assumed that complete combustion occurs and that the basis for the calculations is 100 lb of waste mixture.

The balanced stoichiometric reactions are as follows:

$$C_6H_5Cl + 7\ O_2 \rightarrow CO_2 + 2\ H_2O + HCl$$

$$C_{14}H_9Cl_5 + 15\ O_2 \rightarrow 14\ CO_2 + 2\ H_2O + 5\ HCl$$

The waste mixture analysis is calculated as follows:

chlorobenzene:
0.55 lb CB/lb waste (100 lb waste)/(112.5 lb/lbmol CB) = 0.49 lbmol CB

DDT:
0.45 lb DDT/lb waste (100 lb waste)/(177.5 lb/lbmol DDT) = 0.25 lbmol DDT

Therefore, the total number of moles = 0.49 = 0.25 = 0.74 lbmol

The molar flow rate of dry gas is calculated as follows (100 lb feed basis):

$$\text{lbmol/min dry gas} = 132{,}346\ \text{acfm}\ \frac{(492\ ^\circ R)}{(2500 + 460\ ^\circ R)}\ \frac{1\ \text{lbmol}}{359\ \text{ft}^3}$$

$$= 61.28\ \text{lbmol/min dry gas}$$

The molar flow rate of each component in the gas stream are calculated as follows:

61.28 lbmol/min dry gas (0.106 lbmol CO_2/lbmol dry gas) = 6.5 lbmol/min CO_2
61.28 lbmol/min dry gas (0.094 lbmol O_2/lbmol dry gas) = 5.76 lbmol/min O_2
61.28 lbmol/min dry gas (0.800 lbmol N_2/lbmol dry gas) = 49.0 lbmol/min NO_2

The amount of oxygen required for complete combustion based on the the stoichiometric equations above is:

lbmol O_2 = (7 lbmol O_2/lbmol CB) (0.49 lbmol CB) + (15 lbmol O_2/lbmol DDT) (0.25 lbmol DDT) = **7.18 lbmol O_2**

The amount of excess air in the stack is calculated based on the amount of O_2 in the flue gas versus that required under stoichiometric conditions:

% EA = 5.76 lbmol O_2 in flue gas/7.18 lbmol O_2 stoichiometric requirement
= 0.8022 = **80% EA**

The amounts of HCl and water formed are also based on the balanced equations above:

lbmol HCl formed = (1 lbmol HCl/lbmol CB) (0.49 lbmol CB) + (5 lbmol HCl/lbmol DDT) (0.25 lbmol DDT) = **1.74 lbmol HCl**

lbmol H_2O formed = (2 lbmol H_2O/lbmol CB) (0.49 lbmol CB) + (2 lbmol H_2O/lbmol DDT) (0.25 lbmol DDT) = **1.48 lbmol H_2O**

The total amount of flue gas on a molar basis then becomes:

S-7 Solution (Continued).

Flue gas lbmol = 6.5 lbmol CO_2 + 5.76 lbmol O_2 + 49.02 lbmol N_2 + 1.74 lbmol HCl + 1.48 lbmol H_2O = **64.5 lbmol**

Wet gas acfm = 132,346 dacfm [64.5 lbmol wet gas/(64.5 - 1.48 lbmol dry gas)] = **135,498 acfm**

Flue gas mass flow rate = 6.5 lbmol/min CO_2 (44 lb CO_2/lbmol) + 5.76 lbmol/min O_2 (32 lb O_2/lbmol) + 49.02 lbmol/min N_2 (28 lb N_2/lbmol) + 1.74 lbmol/min HCl (36.45 lb HCl/lbmol) + 1.48 lbmol/min H_2O (18 lb H_2O /lbmol) = **1933 lb/h = 115,977 lb/min**

S-8 Solution.

Neutralization reactions:

$$HCl + NaOH \rightarrow NaCl + H_2O$$

$$SO_2 + 2\ NaOH \rightarrow Na_2SO_3 + H_2O$$

$$CO_2 + NaOH \rightarrow NaHCO_3$$

The standard gas flow rate is:

10,000 acfm (32 + 460 °R)/(180 + 460 °R) = 7700 scfm

The mass rates absorbed are:

HCl: (7700 scfm/359 ft^3/lbmol)(60 min/h)(1500 ppm/10^6) = **1.93 lbmol/h**

SO_2: (7700 scfm/359 ft^3/lbmol)(60 min/h)(470 ppm/10^6) = **0.60 lbmol/h**

CO_2: (7700 scfm/359 ft^3/lbmol)(60 min/h)(0.13 ppm/10^6) = **167 lbmol/h**

Then, the NaOH required is:

0.999 (1.93 lbmol/h) + 0.65 (0.6 lbmol/h) (2) + 0.01 (167 lbmol/h) = **4.38 lbmol/h**

or, 4.38 lbmol/h (40 lb/lbmol)/[1.31 g/mL (8.34 gal/ft^3) (0.28)] = **57 gal/h of 28% pure NaOH solution required**

S-9 Solution.

a. The mass rate of SO_2 and HCl production are obtained as follows:

$PV = nRT$
$PQ = \dot{n}RT$
$\dot{m} = \dot{n}\ (MW) = PQ\ (MW)/RT$

$$\dot{m}_{HCl} = \frac{1\text{ atm }(10{,}000\text{ acfm})\ (0.02\text{ mol fraction HCl})(36.45\text{ lb/lbmol HCl})}{(0.7302\text{ atm ft}^3\text{/lbmol-°R})\ (460 + 750\text{ °R})\ (1\text{ h/60 min})}$$

= **495 lb HCl/h**

S-9 Solution (Continued).

$$\dot{m}_{SO_2} = \frac{1\ \text{atm}\ (10{,}000\ \text{acfm})\ (3.5 \times 10^{-4}\ \text{mole fraction}\ SO_2)\ (64\ \text{lb}\ SO_2/\text{lbmol})}{(0.7302\ \text{atm ft}^3/\text{lbmol-°R})\ (460 + 750\ °\text{R})\ (1\ \text{h}/60\ \text{min})}$$

$= \mathbf{15.2\ lb\ SO_2/h}$

Based on the HCl production rate, it must be determined which HCl discharge standard controls emissions, i.e., 99% removal or 4 lb/h.

Based on the 99% removal requirement, the HCl emission would be: 0.01 (495) = 4.95 lb/h. (Since this limit is more than the 4 lb/h standard, the 99% standard applies).

The balanced chemical equations for the neutralization of acid gases using excess lime are:

$1.3\ Ca(OH)_2 + SO_2 + 0.5\ O_2 \rightarrow CaSO_4 + H_2O + 0.3\ Ca(OH)_2$
$1.1\ Ca(OH)_2 + 2\ HCl \rightarrow CaCl_2 + 2\ H_2O + 0.1\ Ca(OH)_2$

Therefore, SO_2 removal requires 1.3 mol lime/mol SO_2, while HCl removal requires 0.55 mol lime/mol HCl.

The amounts of each acid gas to be removed in the scrubbing process are as follows:

SO_2: (0.7) 15.2 lb/h/(64 lb SO_2/lbmol) = **0.166 lbmol SO_2/h**
HCl: (495 - 4.95 lb HCl/h)/(36.45 lb HCl/lbmol) = **13.6 lbmol HCl/h**

The required lime feed rate is then calculated as:

Lime feed rate = 0.166 lbmol SO_2/h (1.3 lbmol lime/lbmol SO_2) + 13.6 lbmol HCl/h (0.55 lbmol lime/lbmol HCl) = **7.7 lbmol lime/h**

b. The mass production rate of spent solids is:

Dry solids production rate = $\dot{m}_{CaSO_4} + \dot{m}_{CaCl_2} + \dot{m}_{Ca(OH)_2\,(unreacted)}$

= (0.166 lbmol $CaSO_4$/h) (136.14 lb $CaSO_4$/lbmol) + (13.6/2 lbmol $CaCl_2$) (110.9 lb $CaCl_2$/lbmol) + [0.3 (0.166) lbmol $Ca(OH)_2$ {for the SO_2} + 0.1/2 (13.6) lbmol $Ca(OH)_2$ {for the HCl}] (74.1 lb $Ca(OH)_2$/lbmol)
= 32.4 lb $CaSO_4$/h + 754.1 lb $CaCl_2$/h + 55.7 lb $Ca(OH)_2$/h
= **842.2 lb solids/h**

S-10 Solution.

a. The amount of ash leaving the stack is:

$$\frac{0.08\ \text{gr}}{1\ \text{dscf}}\left(\frac{1\ \text{lb}}{7000\ \text{gr}}\right)\left(\frac{20{,}000\ \text{dscf}}{1\ \text{min}}\right)\left(\frac{60\ \text{min}}{1\ \text{h}}\right)\left(\frac{24\ \text{h}}{1\ \text{d}}\right) = \mathbf{329\ lb/d}$$

The amount of ash collected in the ESP is:

(329 lb/d)/(1 - 0.995 collected) = **65,800 lb/d**

S-10 Solution (Continued).

b. The amount of mercury leaving the stack with the fly ash is:

329 lb/d (2.42 x 10^{-6} g Hg/g ash) = **7.96 x 10^{-4} lb/d = 0.361 g/d**

The amount of mercury leaving the stack as vapor is:

0.3 x 10^{-3} g Hg/dscm (20,000 dscf/min) (1 m^3/35.3 ft^3) (60 min/1 h) (24 h/1 d) = **245 g/d**

Total mercury leaving the stack = 245 + 0.361 = **241.4 g/d**

CHAPTER FOUR

Thermochemistry

T-1 Solution.

The balanced stoichiometric equations for the combined chloroethylene and propane stream are as follows:

$$C_2H_3Cl + 5/2\ O_2 + 9.4\ N_2 \rightarrow 2\ CO_2 + H_2O + HCl + 9.4\ N_2$$

$$C_3H_8 + 5\ O_2 + 18.8\ N_2 \rightarrow 3\ CO_2 + 4\ H_2O + 18.8\ N_2$$

The theoretical adiabatic flame temperature can be determined by assuming that all of the energy released upon combustion is used to heat the combustion products to the flame temperature. The equations describing this assumption are as follows:

$$\Delta H_c^o = -\ \Delta H_p^o$$

$$\Delta H_c^o = \sum \Delta H^o_{f,products} - \sum \Delta H^o_{f,\ reactants}$$

$$\Delta H_p^o = \Delta\alpha\ (T - 298) + \frac{1}{2}\ (\Delta\beta)\ (T - 298^2) + \frac{1}{3}\ (\Delta\gamma)\ (T^3 - 298^3)$$

$$\Delta\alpha = \sum_{products} n\alpha - \sum_{reactants} n\alpha$$, with $\Delta\beta$ and $\Delta\gamma$ are similarly estimated

Standard heat of combustion data for the two compounds are calculated as:

chloroethylene:

$$\Delta H_c^o = 2\ \Delta H_f^o(CO_2) + \Delta H_f^o(H_2O) + \Delta H_f^o(HCl) + 9.4\ \Delta H_f^o(N_2)$$
$$-\ \Delta H_f^o(C_2H_3Cl) - 5.2\ \Delta H_f^o(O_2) - 9.4\ \Delta H_f^o(N_2)$$

propane:

$$\Delta H_c^o = 3\ \Delta H_f^o(CO_2) + 4\ \Delta H_f^o(H_2O) + 18.8\ \Delta H_f^o(N_2)$$
$$-\ \Delta H_f^o(C_3H_8) - 5\ \Delta H_f^o(O_2) - 18.8\ \Delta H_f^o(N_2)$$

Based on the data presented in Appendix H for heats of formation, the heats of combustion for these compounds can be calculated as follows:

chloroethylene:

Heat of combustion = 2 (-94,052 cal/gmol) + (-57,798 cal/gmol) + (-22,063 cal/gmol) - (-8400 cal/gmol) = **-259,565 cal/gmol**

propane:

Heat of combustion = 3 (-94,052 cal/gmol) + 4 (-57,798 cal/gmol) - (-24,820 cal/gmol) = **-488,528 cal/gmol**

T-1 Solution (Continued).

Heat capacity values required for the estimation of the final temperature of the product gas (adiabatic flame temperature) are used to calculate values of α, β and γ as shown below, using data from Appendix E. These calculations are based on irreversible reactions where no reactant remains.

chloroethylene:

$\Delta\alpha = 2\ (6.214) + 7.256 + 6.732 + 9.4\ (6.524) = \mathbf{87.7416}$
$\Delta\beta = 2\ (10.396) + 2.298 + 0.433 + 9.4\ (1.250) = \mathbf{35.273 \times 10^{-3}}$
$\Delta\gamma = 2\ (-3.545) + 0.283 + 0.370 + 9.4\ (-0.001) = \mathbf{-6.446 \times 10^{-6}}$

propane:

$\Delta\alpha = 3\ (6.214) + 4\ (7.256) + 18.8\ (6.524) = \mathbf{170.317}$
$\Delta\beta = 3\ (10.396) + 4\ (2.298) + 18.8\ (1.250) = \mathbf{63.820 \times 10^{-3}}$
$\Delta\gamma = 3\ (-3.545) + 4\ (0.283) + 18.8\ (-0.001) = \mathbf{-9.5218 \times 10^{-6}}$

With a mixture of compounds, the mole fraction of each compound in the mixture must be used to determine the flame temperature of the mixture. Mole fraction calculations are made as follows:

MW chloroethylene = 63.5 lb/lbmol; MW propane = 44 lb/lbmol

Assuming 100 lb of the mixture, the number of moles of chloroethylene = 75 lb/62.5 lb/lbmol = 1.2 lbmol, while the number of moles of propane = 25/44 = 0.57 lbmol. Therefore the total number of moles = 1.2 + 0.57 = 1.77 lbmol. The mole fraction of each compound is then:

chloroethylene = 1.2/1.77 = **0.679**
propane = 0.57/1.77 = **0.321**

Assume 1 gmol of the mixture as a basis for the calculations:

$$\Delta H_p^\circ = (T - 298)\ [\ 0.679\ (87.7416) + 0.321\ (170.317)]$$
$$(T^2 - 298^2)/2\ [\ 0.679\ (35.273 \times 10^{-3}) + 0.321\ (63.82 \times 10^{-3})]$$
$$(T^3 - 298^3)/3\ [\ 0.679\ (-6.446 \times 10^{-3}) + 0.321\ (-9.5218 \times 10^{-6})]$$
$$\Delta H_p^\circ = 114.25\ (T - 298) + 44.44 \times 10^{-3}\ (T^2 - 298^2) - 7.434 \times 10^{-6}\ (T^3 - 298^3)$$

$\Delta H_c^\circ = 0.679\ (-259{,}565) + 0.321\ (-488{,}528) =$ **-333,032 cal/gmol**

Equating the heat of combustion to the negative change in enthalpy of the products (from 298 K to the adiabatic flame temperature) yields the following:

$$-2.478 \times 10^{-6}\ T^3 + 0.02222\ T^2 + 114.25\ T - 396{,}016 = 0$$

Solving this expression assuming the cubic term can be neglected yields a flame temperature = 2372 K. Solving the equation more rigorously using Newton's method or spreadsheet iteration yields an adiabatic flame temperature = **2406 K = 3871 °F.**

T-2 Solution.

For the stoichiometric combustion of this waste, the following reactions are assumed. (All quantities are based on 100 g of waste).

$C + O_2 \rightarrow CO_2$; 1 mol O_2 per mol C; 35 g/(12 g/gmol C) = 2.92 gmol C
2 H + 1/2 $O_2 \rightarrow H_2O$; 1/4 mol O_2 per mol H; 7 g/(1 g/gmol H) = 7 gmol H
$S + O_2 \rightarrow SO_2$; 1 mol O_2 per mol S; 2 g/(32.1 g/gmol S) = 0.062 gmol S
$CH_4 + 2\ O_2 \rightarrow CO_2 + 2\ H_2O$; 2 mol O_2 per mol CH_4; 100 g/(16 g/gmol CH_4) = 6.25 gmol CH_4

The gmol of O_2 required for the stoichiometric combustion of the waste/CH_4 mixture is:

gmol O_2 = 2.92 gmol C (1 mol/mol) + 7 gmol H (0.25 mol/mol) + 0.062 gmol S (1 mol/mol) - 11 g O/16 g/gmol O + 6.25 gmol (2 mol/mol) = 16.54 gmol O_2 required

For 100% EA, the required oxygen = 2 x stoichiometric = **33.08 gmol O_2 required**

Amount of nitrogen entering the incinerator with oxygen is:

N_2 = (33.08 gmol) (79/21) = **124.44 gmol N_2**

Total amount of the product flue gas formed is:

CO_2: 2.92 gmol (CO_2) + 6.25 gmol (CH_4) = 9.17 gmol
H_2O: 45 g/18 g/gmol (waste) + 7/2 gmol (H_2O) + 6.25 (2) gmol (CH_4) = 18.5 gmol
SO_2: 0.0625 gmol (SO_2) = 0.0625 gmol
N_2: 124.44 gmol
O_2: 16.54 gmol

Based on the data presented in Appendix H for the net heating value for these compounds, the total heat of combustion for the mixture can be calculated as follows:

$\Delta H_c°$ = 2.92 gmol C (-94,052 cal/gmol C) + 7/2 gmol H_2 (-57,798 cal/gmol H_2) + 0.062 gmol S (-70,960 cal/gmol S) - 45 g H_2O/18 g H_2O/gmol (-10,519 cal/gmol H_2O) + 6.25 gmol CH_4 (-191,762 cal/gmol CH_4) = **-1,653,540 cal/gmol**

Heat capacity values required for the estimation of the final temperature of the product gas (adiabatic flame temperature) are used to calculate values of α, β and γ as shown below, using data from Appendix E. These calculations are based on irreversible reactions where no reactant remains.

$\Delta\alpha$ = 9.17 (6.214) + 18.5 (7.256) + 0.0625 (7.116) + 124.44 (6.524) + 16.54 (6.148) = **1105.2**
$\Delta\beta$ = 9.17 (10.396) + 18.5 (2.298) + 0.0625 (9.512) + 124.44 (1.250) + 16.54 (3.102) = **345.3 x 10^{-3}**
$\Delta\gamma$ = 9.17 (-3.545) + 18.5 (0.283) + 0.0625 (3.511) + 124.44 (-0.001) + 16.54 (-0.923) = **-42.44 x 10^{-6}**

Therefore,

$\Delta H_p°$ = (T - 298) (1105.2) + [(T^2 - 298^2)/2] (345.3 x 10^{-3}) + [(T^3- 298^3)/3] (-42.44 x 10^{-6})
$\Delta H_p°$ = 1105.2 (T - 298) + 172.65 x 10^{-3} (T^2 - 298^2) - 14.147 x 10^{-6} (T^3 - 298^3)

T-2 Solution (Continued).

Equating heat of combustion to the negative change in enthalpy of the products (from 298 K to the adiabatic flame temperature) yields the following:

$$-14.147 \times 10^{-6}\ T^3 + 0.1726\ T^2 + 1105.2\ T - 2{,}327{,}197 = 0$$

Solving the equation rigorously using a spreadsheet iteration procedure yields an adiabatic flame temperature = **1712 K = 2622 °F**.

T-3 Solution.

a. For pure oxygen combustion, the stoichiometric expression is:

$$CCl_4 + O_2 \rightarrow CO_2 + 2\ Cl_2$$

The standard heat of combustion data are obtained from the heats of formation for these compounds from Appendix H.

$$\Delta H^\circ_c = \Delta H^\circ_f\ (CO_2) - \Delta H^\circ_f\ (CCl_4) = -94{,}052\ \text{cal/gmol} + 24{,}000\ \text{cal/gmol}$$
$$= \mathbf{-70{,}052\ cal/gmol}$$

The heat capacity of the flue gas products is given as follows based on data presented in Appendix F:

$$\Delta C_p = C_p\ (CO_2) + 2\ C_p\ (Cl_2) \quad (\text{cal/gmol-K})$$
$$= [10.57 + 2.1 \times 10^{-3}\ (T) - 2.06 \times 10^5\ (1/T^2)]$$
$$+ 2\ [8.85 + 0.16 \times 10^{-3}\ (T) - 0.68 \times 10^5\ (1/T^2)]$$
$$= 28.27 + 2.42 \times 10^{-3}\ (T) - 2.74 \times 10^5\ (1/T^2)$$

The enthalpy change associated with heating the flue gas products is related to the integrated heat capacity of these products from standard temperature (298 K) to the final adiabatic flame temperature (T):

$$\Delta H_p^\circ = (T - 298)\ (28.27) + [(T^2 - 298^2)/2]\ (2.42 \times 10^{-3})$$
$$+ [(1/T) - (1/298)]\ (2.74 \times 10^5)$$

$$\Delta H_p^\circ = 1.21 \times 10^{-3}\ T^2 + 28.27\ T + 2.74 \times 10^5/T - 7397.55$$

Equating the enthalpy change of the flue gas to the negative of the heat of combustion and solving yields the following equation:

$$1.21 \times 10^{-3}\ T^2 + 28.27\ T + 2.74 \times 10^5/T - 77{,}449.55 = 0$$

Trial and error solution of this equations yields a value of T = **2473.8 K = 3993 °F**

b. For the theoretical adiabatic flame temperature calculation, the stoichiometric expression is:

$$CCl_4 + O_2 + 3.76\ N_2 \rightarrow CO_2 + 2\ Cl_2 + 3.76\ N_2$$

The standard heat of combustion data are obtained from the heats of formation for these compounds from Appendix H.

$$\Delta H^\circ_c = \Delta H^\circ_f\ (CO_2) - \Delta H^\circ_f\ (CCl_4) = -94{,}052\ \text{cal/gmol} + 24{,}000\ \text{cal/gmol}$$
$$= \mathbf{-70{,}052\ cal/gmol}$$

The heat capacity of the flue gas products is given as follows based on data presented in Appendix F:

T-3 Solution (Continued).

$$\Delta C_p = C_p\,(CO_2) + 2\,C_p\,(Cl_2) + 3.76\,C_p\,(N_2) \quad \text{(cal/gmol-K)}$$
$$= [10.57 + 2.1 \times 10^{-3}\,(T) - 2.06 \times 10^5\,(1/T^2)]$$
$$+ 2\,[8.85 + 0.16 \times 10^{-3}\,(T) - 0.68 \times 10^5\,(1/T^2)]$$
$$+ 3.76\,[6.83 + 0.9 \times 10^{-3}\,(T) - 0.12 \times 10^5\,(1/T^2)]$$
$$= 53.95 + 5.8 \times 10^{-3}\,(T) - 3.87 \times 10^5\,(1/T^2)$$

The enthalpy change associated with heating the flue gas products is related to the integrated heat capacity of these products from standard temperature (298 K) to the final adiabatic flame temperature (T):

$$\Delta H_p^\circ = (T - 298)\,(53.95) + [(T^2 - 298^2)/2]\,(5.8 \times 10^{-3}) + [(1/T) - (1/298)]\,(3.87 \times 10^5) = 2.9 \times 10^{-3}\,T^2 + 53.95\,T + 3.87 \times 10^5/T - 17633.3$$

Equating the enthalpy change of the flue gas to negative of the heat of combustion and solving yields the following equation:

$$2.9 \times 10^{-3}\,T^2 + 53.95\,T + 3.87 \times 10^5/T - 87{,}685.3 = 0$$

Trial and error solution of this equation yields a value of T = **1334.5 K = 1942 °F.**

c. For the 100% EA flame temperature calculation, the stoichiometric expression is:

$$CCl_4 + 2\,O_2 + 7.52\,N_2 \rightarrow CO_2 + 2\,Cl_2 + O_2 + 7.52\,N_2$$

The standard heat of combustion data are obtained from the heats of formation for these compounds from Appendix H.

$$\Delta H^\circ_c = \Delta H^\circ_f(CO_2) - \Delta H^\circ_f(CCl_4) = -94{,}052\text{ cal/gmol} + 24{,}000\text{ cal/gmol}$$
$$= \mathbf{-70{,}052\ cal/gmol}$$

The heat capacity of the flue gas products is given as follows based on data presented in Appendix F:

$$\Delta C_p = C_p\,(CO_2) + 2\,C_p\,(Cl_2) + 7.52\,C_p\,(N_2) + C_p\,(O_2) \quad \text{(cal/gmol-K)}$$
$$= [10.57 + 2.1 \times 10^{-3}\,(T) - 2.06 \times 10^5\,(1/T^2)]$$
$$+ 2\,[8.85 + 0.16 \times 10^{-3}\,(T) - 0.68 \times 10^5\,(1/T^2)]$$
$$+ 7.52\,[6.83 + 0.9 \times 10^{-3}\,(T) - 0.12 \times 10^5\,(1/T^2)]$$
$$+ [7.16 + 1.0 \times 10^{-3}\,(T) - 0.4 \times 10^5\,(1/T^2)]$$
$$= 86.79 + 10.19 \times 10^{-3}\,(T) - 4.72 \times 10^5\,(1/T^2)$$

The enthalpy change associated with heating the flue gas products is related to the integrated heat capacity of these products from standard temperature (298 K) to the final adiabatic flame temperature (T):

$$\Delta H_p^\circ = (T - 298)\,(86.79) + [(T^2 - 298^2)/2]\,(10.19 \times 10^{-3}) + [(1/T) - (1/298)]\,(4.72 \times 10^5)$$

$$\Delta H_p^\circ = 5.095 \times 10^{-3}\,T^2 + 86.79\,T + 4.72 \times 10^5/T - 27899.75$$

Equating the enthalpy change of the flue gas to negative of the heat of combustion and solving yields the following equation:

$$5.095 \times 10^{-3}\,T^2 + 86.79\,T + 4.72 \times 10^5/T - 97951.75 = 0$$

Trial and error solution of this equations yields a value of T = **1057.8 K = 1444 °F.**

T-4 Solution.

The waste mixture contains 30% cellulose, 30% motor oil, 20% water and 20% inerts. The NHV of the waste mixture is approximated by:

$NHV_{mixture}$ = [0.3 (14,000 Btu/lb cellulose) + 0.3 (25,000 Btu/lb motor oil) + 0.2 (0 Btu/lb water) + 0.2 (-1,000 Btu/lb inerts)] (1 - 0.05 heat loss) = **10,925 Btu/lb mixture**

The theoretical flame temperature can be estimated using the Theodore-Reynolds equation, and the simplified relationship given for the determination of EA as follows for 10% excess air:

The excess air fraction = [0.95 (10.4)]/(21 - 10.4) = 0.932

Therefore, the adiabatic flame temperature can be calculated as:

$$T = 60 + \frac{10{,}925}{(0.325)[1 + (1.932)(7.5 \times 10^{-4})(10{,}925)]}$$

$$T = \mathbf{1{,}997\ ^{\circ}F}$$

T-5 Solution.

a. Assume a working basis of 100 lb/h of waste mixture containing 20 lb/h CCl_4 and 80 lb/h CH_4.

The componential composition of the waste is:

	Mass, %	Mole, %
CCl_4	20	2.54
CH_4	80	97.46

The elemental composition in the waste is:

	lb/h	Mass, %
C	61	61.5
H	20	20.1
Cl	18	18.4

The following equations are written to describe the combustion reactions taking place within the incinerator:

$$CCl_4 + O_2 \rightarrow CO_2 + 2\ Cl_2 \quad (1)$$

$$CH_4 + 2\ O_2 \rightarrow CO_2 + 2\ H_2O \quad (2)$$

$$2\ Cl_2 + 2\ H_2O \leftrightarrow 4\ HCl + O_2 \quad (3)$$

Note that at high temperatures, equilibrium favors the formation of HCl.

T-5 Solution (Continued).

The stoichiometric O_2 requirement is :

O_2 required = O_2 (Equation 1) + O_2 (Equation 2) - O_2 (Equation 3)
= 1 lbmol/lbmol CCl_4 + 2 lbmol/lbmol CH_4 - 1 lbmol/lbmol CCl_4

= [(20 lb CCl_4/h)/{(12 + 35.45 (4 lb/lbmol CCl_4)}) + 2 (80 lbmol/h CH_4)/(12 + 4 lb/lbmol CH_4) - (20 lb CCl_4/h)/{12 + 35.54 (4 lb/lbmol CCl_4)}] 32 lb O_2/lbmol
= (0.13 lbmol CCl_4/h + 10 lbmol CH_4/h - 0.13 lbmol CCl_4/h) 32 lb O_2/lbmol
= **320 lb O_2/h**

Therefore, the air requirement = 320 lb O_2/h (4.31 lb air/lb O_2)
= **1379 lb air/h**

The amount of O_2 in flue gas would be zero because the unit is operating at 0 % EA.

b. Adiabatic combustion temperature at 0 % EA can be calculated assuming that all Cl_2 is converted to HCl. The heat of combustion for this reaction is -42,708 cal/gmol.

$$\Delta H_p = \int_{298}^{T} \Delta C_p \, dT \qquad (4)$$

ΔHp = - $n_1[\Delta H_c{}^\circ(CCl_4)]$ - $n_2[\Delta H_c{}^\circ(CH_4)]$
= - 0.0254 (-42,708 cal/gmol) + 0.9746 (212,880 cal/gmol)
= **208,558 cal/gmol**

Composition of the flue gas is given below based on the following stoichiometric equations and relationships:

$$0.13\ CCl_4 + 0.13\ O_2 \rightarrow 0.13\ CO_2 + 0.26\ Cl_2 \qquad (1a)$$

$$5\ CH_4 + 10\ O_2 \rightarrow 5\ CO_2 + 10\ H_2O \qquad (2a)$$

$$0.26\ Cl_2 + 0.26\ H_2O \leftrightarrow 0.52\ HCl + 0.13\ O_2 \qquad (3a)$$

CO_2: 1 lbmol/lbmol CCl_4, 1 lbmol/lbmol CH_4; lbmol CO_2 = 0.13 + 5 = 5.13
H_2O: 2 lbmol/lbmol CH_4, - 2 lbmol/lbmol CCl_4;
lbmol H_2O = 2 (5) - 2 (0.13) = 9.74
HCl: 4 lbmol/lbmol CCl_4, lbmol HCl = 4 (0.13) = 0.52
N_2: 1059 lb/h

	lb/h	Mass %	Mole %
CO_2	225.7	15.3	9.64
H_2O(g)	175.3	11.85	18.3
N_2	1059	71.6	71.1
HCl	18.95	1.29	0.98
O_2 (0% EA)	0	0	0

The heat capacity of the flue gas products is given as follows based on data presented in Appendix F, and the composition of the flue gas above.

T-5 Solution (Continued).

$C_p(CO_2) = 10.57 + (2.1 \times 10^{-3})(T) - 2.06 \times 10^5 (1/T^2)$

$C_p(H_2O) = 7.3 + 2.46 \times 10^{-3} (T)$

$C_p(N_2) = 6.83 + (0.9 \times 10^{-3})(T) - 0.12 \times 10^5 (1/T^2)$

$C_p(HCl) = 6.27 + (1.24 \times 10^{-3})(T) + 0.3 \times 10^5 (1/T^2)$

$C_p = 0.0964\ C_p(CO_2) + 0.183\ C_p(H_2O) + 0.711\ C_p(N_2) + 0.0098\ C_p(HCl)$

The enthalpy change associated with heating the flue gas products is related to the integrated heat capacity of these products from standard temperature (298 K) to the final adiabatic flame temperature (T):

$$\Delta H_p^\circ = (T - 298)(6.359) + [(T^2 - 298^2)/2](1.304 \times 10^{-3}) + (1/T - 1/298)(28{,}682)$$

Equating the enthalpy change of the flue gas to the negative of the heat of combustion and solving by trial and error yields a value of T = **2005 K = 3149 °F.**

c. Excess air required to maintain the operating temperature at 2100 °F is calculated by rewriting Equations 1, 2 and 3 as:

$$CCl_4 + X\ O_2 + 3.76X\ N_2 \rightarrow CO_2 + 2Cl_2 + (X - 1)\ O_2 + 3.76X\ N_2$$

$$CH_4 + 2Y\ O_2 + 7.52Y\ N_2 \rightarrow CO_2 + H_2O + (2Y - 2)\ O_2 + 7.52Y\ N_2$$

$$Cl_2 + H_2O \leftrightarrow 2\ HCl + 1/2\ O_2$$

Where X and Y are fractions of excess air. Similar calculations can be carried out as in Part b. However, the problem was solved using a computer program written by Prof. Reynolds at Manhattan College. Results of the simulation are summarized below.

Flue gas composition

	lb	Mass %	Mole %
CO_2	225	6.99	4.51
H_2O(g)	175	5.43	8.55
N_2	2395	74.32	75.26
O_2	408	12.66	11.22
HCl	19	0.59	0.46
Cl_2	0	0	0

It was found that 128% excess air is required to maintain the operating temperature at 2100 °F.

T-6 Solution.

The following data are provided from Appendix H:

$\Delta H^{\circ}_{c\,C_6H_6(g)}$ = -789,080 cal/gmol
$\Delta H^{\circ}_{c\,C}$ = -94,052 cal/gmol
$\Delta H^{\circ}_{c\,H_2}$ = -68,317 cal/gmol

The heat of combustion of six moles of carbon and three moles of hydrogen molecules is:

ΔH_c = 6 (-94,052 cal/gmol) + 3 (-68,317 cal/gmol) = **-769,263 cal/gmol**

The error involved in such a calculation is:

% Error = [-798,080 - (-769,263)]/(-789,080) = 0.025 = **2.5%**

T-7 Solution.

The stoichiometry for the problem is as follows:

$$C_2H_5OH + 3\ O_2 \rightarrow 2\ CO_2 + 3\ H_2O$$

Assuming a 1 lb waste basis, the stoichiometric O_2 = (0.5)(3 lbmol) (32 lb/lbmol O_2)/(46 lb/lbmol EtOH) = **1.04 lb O_2**

Then, the oxygen requirement at 40% EA = **1.46 lb O_2/lb waste**

The composition of the flue gas is given as:

N_2: (1.46 lb O_2/lb waste [(28 lb/lbmol N_2)/(32 lb/lbmol O_2)] (71 lbmol N_2/29 lbmol O_2) = **3.13 lb N_2/lb waste**

CO_2: (0.5 lb EtOH/lb waste) [2 (44 lb CO_2/lbmol)/(46 lb EtOH/lbmol)] = **1.91 lb CO_2/lb waste**

O_2: (0.4 EA)(1.04) = **0.42 lb O_2/lb waste**

H_2O: (0.5 lb H_2O/lb waste) + (0.5 lb EtOH/waste) [3 (18 lb H_2O/lbmol)/(46 lb EtOH/lbmol)] = **1.09 lb H_2O/lb waste, 0.5 lb of which comes in with waste.**

Total N_2 + CO_2 + O_2 in the dry flue gas = 3.13 + 1.91 + 0.42 = **5.46 lb/lb waste**

The NHV for ethyl alcohol from Appendix G = **11,929 Btu/lb**

The NHV for the combined water/ethyl alcohol waste is:

NHV = 0.5 lb EtOH/lb waste (11,929 Btu/lb EtOH)
- 0.5 lb H_2O/lb waste (1 Btu/lb H_2O-°F) (212 - 60) (to heat water to boiling) - 0.5 lb H_2O/lb waste (970 Btu/lb H_2O) (latent heat of vaporization) - 0.5 lb H_2O/lb waste (0.5 Btu/lb H_2O-°F) (T - 212) (to raise water vapor to final flame temperature)
= 5403.5 Btu/lb waste - 0.25 (T - 212) = **5456.5 - 0.25 T Btu/lb waste**

The energy of the waste is used to raise the temperature of the combustion products, and this change in enthalpy of the combustion products is calculated as:

T-7 Solution (Continued).

ΔH_p = 5.46 lb gas/lb waste (0.26 Btu/lb gas-°F) (T - 60) + 0.59 lb H_2O/lb waste (1 Btu/lb H_2O-°F) (212 - 60) + 0.59 lb H_2O/lb waste (970 Btu/lb H_2O for latent heat of vaporization) + 0.59 lb H_2O/lb waste (0.5 Btu/lb H_2O-°F) (T - 212)

= 1.42 (T - 60) + 89.68 + 572.3 + 0.295 (T - 212) = 1.715 T + 514.24

Equating NHV to ΔH_p yields the following relationship and solution:

5456.5 - 0.25 T = 1.715 T + 514.24

4942.26 = 1.965 T; T = 2515.1 = **2,515 °F**

T-8 Solution.

a. Carbon tetrachloride combustion at 2400 °F with 30% excess air is considered on a 1 lbmol CCl_4 basis. The general stoichiometric equation is as follows:

$$n\, CH_4 + CCl_4 + 1.3\,(1+n)\,[O_2 + 3.76\, N_2] \rightarrow (1+n)\, CO_2 + 4\,n\, HCl + (n-1)\, H_2O + 2\,(1-n)\, Cl_2 + 0.3\,(1+n)\, O_2 + 3.76\,(1.3)\,(1+n)\, N_2$$

Two special cases arise depending upon the value of n:

1. n< 1.0 where no water is formed, and
2. n> 1.0 where no chlorine is formed.

The value of n is estimated using the NHV of the chlorinated hydrocarbon and fuel using DuLong's equation for individual NHV values:

$$NHV = NVH(CCl_4) + n\, NVH(CH_4)$$

DuLong's Equation:

NHV≈ 14,000 mass%C + 45,000 (mass%H - 1/8 mass%O) - 760 mass%Cl + 4500 mass%S

For CCl_4:

mass%C = 12/[12 + 4 (35.45)] = 0.078
mass%Cl = 1 - mass%C = 0.921

NHV(CCl_4) = 14,000 (0.078) - 760 (0.921) = **392 Btu/lb**

For CH_4:

mass%C = 12/(12 + 4) = 0.75
mass%H = 1 - mass%C = 0.25

NHV(CH_4) = 14,000 (0.75) + 45,000 (0.25) = **21,750 Btu/lb**

Therefore, the NHV of the carbon tet/fuel mixture is:

NHV = 392 + n (21,750) Btu/lb mixture (1)

The NHV can be related to incinerator temperature and excess oxygen through the following equation:

T-8 Solution (Continued).

$$NHV = \frac{0.3\,(T - 60)}{[1 - (1 + EA)(0.3)(7.5 \times 10^{-4})(T - 60)]} \quad (2)$$

Equating Equations 1 and 2 above to one another with EA = 0.3 and T = 2400 yields the following expression:

NHV = 2225 = 392 + n (21,750); **n = 0.084**

Thus n < 1.0, and Cl_2 will be formed while no water is produced. The resultant stoichiometric expression is as follows:

$$0.084\ CH_4 + CCl_4 + 1.409\ [O_2 + 3.76\ N_2] \rightarrow$$
$$(1.084)\ CO_2 + 0.336\ HCl + 1.832\ Cl_2 + 0.3252\ O_2 + 5.299\ N_2$$

b. The weight% of each component of the gas leaving the unit is calculated based on the balanced equation presented above as:

CO_2: 1.084 mol (44 lb/lbmol) = 47.7 lb
HCl: 0.336 mol (36.45 lb/lbmol) = 12.25 lb
Cl_2: 1.832 mol (70.9 lb/lbmol) = 129.9 lb
O_2: 0.3252 mol (32 lb/lbmol) = 10.41 lb
N_2: 5.299 mol (28 lb/lbmol) = 148.37 lb

Total mass = 348.63 lb flue gases

The mass% of each flue gas component is then determined to be as follows:

CO_2: 47.7 lb/348.6 lb = 0.137 = 13.7%
HCl: 12.25 lb/348.6 lb = 0.035 = 3.5%
Cl_2: 129.9 lb/348.6 lb = 0.373 = 37.3%
O_2: 10.41 lb/348.6 lb = 0.03 = 3.0%
N_2: 148.37 lb/348.6 lb = 0.426 = 42.6%

T-9 Solution.

The constant of integration is obtained from the known value of K at 298 K, which can be obtained from free energies of formation.

$\Delta G° = \Delta G°_f(H_2O) - 2\ [\Delta G°_f(HBr)] = -29,195$ cal/gmol

Then, from $\ln(K) = -\Delta G°/RT = 49.3$, an equilibrium constant can be calculated: **$K = 2.576 \times 10^{21}$**

Thus, formation of Br_2 is heavily favored at low temperatures.

Similarly, $\Delta H°$ is obtained from enthalpies of formation:

$\Delta H° = \Delta H°_f(H_2O) - 2\ [\Delta H°_f(HBr)] = -40,478$ cal/gmol = $-72,860$ Btu/lbmol

Then, ΔC_p is obtained from the average heat capacity values given in the problem and in Appendix C:

$\Delta C_p = C_p(Br_2) + C_p(H_2O) - 2\ C_p(HBr) - .5\ C_p(O_2)$

$$\mathbf{\Delta C_p = -1.583\ Btu/lbmol\text{-}°R}$$

T-9 Solution (Continued).

Using these values of K, ΔH°, and ΔC_p, the constant of integration is evaluated using the following equation:

$$49.3 = \frac{72{,}860 \text{ Btu/lbmol} + 1.538 \text{ Btu/lbmol-}^\circ\text{R } (536\ ^\circ\text{R})}{R\ (536\ ^\circ\text{R})}$$

$$- \frac{1.538 \text{ Btu/lbmol-}^\circ\text{R } (\ln(536\ ^\circ\text{R}))}{R} + I$$

Thus, **I = - 14.89**

Then, at 4000 °F, the same equation is used to evaluate K:

$$\ln(K) = \frac{72{,}860 \text{ Btu/lbmol} + 1.583 \text{ Btu/lbmol-}^\circ\text{R } (536\ ^\circ\text{R})}{R\ (4460\ ^\circ\text{R})}$$

$$- \frac{1.583 \text{ Btu/lbmol-}^\circ\text{R } (\ln(4460\ ^\circ\text{R}))}{R} - 14.89$$

Therefore, **$K = 1.73 \times 10^{-6}$**

An equilibrium constant less than 1 means that the partial pressures of the starting materials will be greater than those of the products for the equilibrium equation as written. Thus, Br_2 levels should be greatly reduced at this higher temperature.

T-10 Solution.

a. F can be defined as the fraction of sludge utilized in the mixture; thus (1 - F) is the fraction of fuel oil used. To obtain a minimum of 8000 Btu/lb:

6000 Btu/lb F + 15,000 Btu/lb (1-F) = 8000 Btu/lb; **F = 0.778**

The amount of sludge that can be burned at the maximum heat rate is:

0.778 (40 x 10^6 Btu/h) (1 lb/8000 Btu) = **3890 lb/h**

The amount of sludge that can be burned at the minimum heat rate is:

0.778 (25 x 10^6 Btu/h) (1 lb/8000 Btu) = **2431 lb/h**

No blending is necessary for the plating waste, so the amount of plating waste that can be incinerated at the maximum heat rate is:

(40 x 10^6 Btu/h) (1 lb/8000 Btu) = **5000 lb/h**

The amount of plating waste that can be burned at the minimum heat rate is:

(25 x 10^6 Btu/h) (1 lb/8000 Btu) = **3125 lb/h**

T-10 Solution (Continued).

Let Q_{plate} and Q_{sludge} represent the flow rate of the two wastes to the hazardous waste incinerator in lb/h. The restrictions on the available blending options are represented on the figure shown below using numbers for each line (constraint). The arrows represent the feasible region with respect to each line.

Line 1: limit of 5000 lb/h of plating waste	$Q_{plate} < 5000$
Line 2: min. of 1000 lb/h of plating waste	$Q_{plate} > 1000$
Line 3: limit of 5000 lb/h of sludge	$Q_{sludge} < 5000$
Line 4: min. of 1000 lb/h of sludge	$Q_{sludge} > 1000$
Line 5: max. heat rate of 40 MBtu/h	$Q_{sludge} = 3890-(3890/5000)\ Q_{plate}$
Line 6: min. heat rate of 25 MBtu/h	$Q_{sludge} = 2431-(2431/3125)\ Q_{plate}$

Lines 5 and 6 are determined by relaxing all constraints and assuming that only sludge is to be incinerated.

Point A represents the maximum rate at which plating waste can be burned. This occurs where Constraints 4 and 5 intersect, i.e.:

1000 lb/h = 3890 lb/h - (3890 lb/h)/(5000 lb/h) Q_{plate}; Q_{plate} = **3715 lb/h**

b. Point B represents the maximum rate at which the sludge can be burned. This occurs where Constraints 2 and 5 intersect, i.e.:

Q_{sludge} = 3890 lb/h -(3890 lb/h)/(5000 lb/h) 1000 = **3112 lb/h**

To determine the fuel oil requirements at this sludge loading, the NHV of the waste is first calculated as:

[(3112 lb/h)/(3112 lb/h + 1000 lb/h)](6000 Btu/lb) + [(1000 lb/h)/(3112 lb/h + 1000 lb/h)](8000 Btu/lb) = **6420 Btu/lb**

Then the fraction of the fuel, (1-F), required to bring the mixture to 8000 Btu/lb is calculated as:

6420 Btu/lb F + 15,000 Btu/lb (1-F) = 8000 Btu/lb; **F = 0.18**

The supplemental fuel requirement at 40 MBtu/h is:

0.18 (40 x 10^6 Btu/h) (1 lb/15,000 Btu) = **480 lb/h**

c. A seventh constraint is required to keep the Cd < 150 lb/h, i.e.:

0.02 Q_{sludge} + 0.08 Q_{plate} < 150 lb/h OR Q_{sludge} = 7500 - 4 Q_{plate}

This constraint in shown as Line 7 on the figure. Given all seven constraints, Point C on the figure permits the maximum incineration rate of the plating waste. This occurs where Constraints 6 and 7 intersect, i.e.:

7500 lb/h - 4 Q_{plate} = 2431 lb/h - (2431 lb/h)/(3125 lb/h) Q_{plate}

Q_{plate} = (7500 lb/h - 2431 lb/h)/(4 - 2431 lb/h)/(3125 lb/h) = **1573 lb/h**

T-10 Solution (Continued).

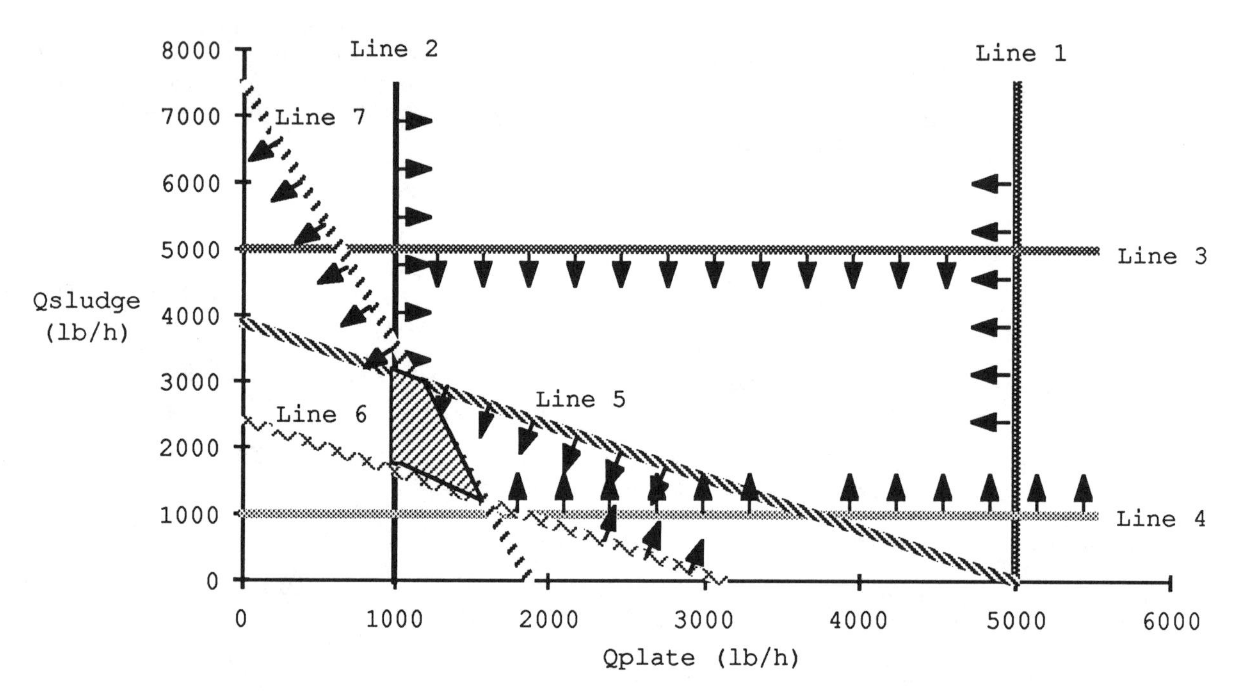

d. The shaded area on the figure represents the permissible operating range for this unit given all constraints. The operating range does not allow much flexibility. The solution is deterministic, and variation in waste, operator error, waste analysis uncertainty, etc., are unaccounted for. Low intensity burners may allow a decrease in the required NHV of the waste and permit additional flexibility.

This graphical technique is adequate for two variables investigated in this problem. This problem can be solved using optimization techniques such as linear programming. This technique can handle numerous constraints and variables, and can be used, for example, to minimize supplemental fuel costs while operating within given system constraints.

T-11 Solution.

The first step in the solution of this problem is to write the balanced oxidation reaction equation for each compound, and from these balanced equations, calculate the standard heat of combustion as follows:

CH_4: $CH_4 + 2\ O_2 \rightarrow CO_2 + 2\ H_2O(g)$
$CHCl_3$: $CHCl_3 + 7/6\ O_2 \rightarrow CO_2 + 1/3\ H_2O + 1/3\ HCl + 4/3\ Cl_2$
C_6H_6: $C_6H_6 + 15/2\ O_2 \rightarrow 6\ CO_2 + 3\ H_2O(g)$
C_6H_5Cl: $C_6H_5Cl + 7\ O_2 \rightarrow CO_2 + 2\ H_2O(g) + HCl$
H_2S: $H_2S + 3/2\ O_2 \rightarrow SO_2 + H_2O(g)$

For the methane reaction, the following heat of combustion calculation can be made:

$$\Delta Hc^\circ = (-94{,}052\ \text{cal/gmol } CO_2) + 2\ (-57{,}798\ \text{cal/gmol } H_2O(g)) - (-17{,}889\ \text{cal/gmol } CH_4) = \mathbf{-191{,}759\ cal/gmol}$$

T-11 Solution (Continued).

Heat of combustion values for the balance of these compounds are calculated in a similar manner. The results of these calculations are summarized in the table below using the conversion 1.8 (cal/g) = Btu/lb.

Compound	NHV (gal/gmol)	MW (g/gmol)	NHV (cal/g)	NHV (Btu/lb)
CH_4	-191,759	16	-11,985	-21,593
$CHCl_3$	-96,472	119.5	-807	-1453
C_6H_6	-717,886	78	-9204	-16,583
C_6H_5Cl	-714,361	112.5	-6350	-11,441
H_2S	-123,943	34	-3645	-6567

Dulong's equation can be written as follows:

NHV≈ 14,000 mass%C + 45,000 (mass%H - 1/8 mass%O) - 760 (mass%Cl) + 4500 (mass%S)

with NHV in the above equation in Btu/lb. The following results can be summarized as:

Compound	mass%C	mass%H	mass%O	mass%Cl	mass%S	NHV (Btu/lb)
CH_4	0.75	0.25	0	0	0	21,750
$CHCl_3$	0.10	0.0084	0	0.891	0	1101
C_6H_6	0.92	0.08	0	0	0	16,480
C_6H_5Cl	0.64	0.04	0	0.32	0	10,520
H_2S	0	0.06	0	0	0.94	6,930

Based on these calculations, the differences between the thermodynamically based NHV values, and those estimated using Dulong's equation can be summarized as:

Compound	NHV_{thermo} (Btu/lb)	NHV_{Dulong} (Btu/lb)	% Difference
CH_4	21,593	21,750	0.73
$CHCl_3$	1180	1101	6.7
C_6H_6	16,583	16,480	0.62
C_6H_5Cl	11,441	10,520	8.05
H_2S	6567	6930	5.53

T-12 Solution.

The species to be considered in the problem, and their respective enthalpies of formation are as follows:

$CH_4(g)$: -17.889 kcal/gmol, $CH_3Cl(l)$: -20.63 kcal/gmol, $CH_2Cl_2(l)$: -22.8 kcal/gmol, $CHCl_3(l)$: -24.2 kcal/gmol, and $CCl_4(l)$: -24.0 kcal/gmol.

For the methane reaction, the following stoichiometric equation and heat of combustion calculation can be written:

CH_4: $CH_4 + 2\ O_2 \rightarrow CO_2 + 2\ H_2O(g)$

ΔH_c° = (-94,052 cal/gmol CO_2) + 2 (-57,798 cal/gmol $H_2O(g)$) - (-17,889 cal/gmol CH_4) = **-191,759 cal/gmol**

T-12 Solution (Continued).

For the methyl chloride reaction:

CH_3Cl: $CH_3Cl + 3/2\ O_2 \rightarrow CO_2 + HCl + H_2O(g)$

$\Delta Hc°$ = (-94,052 cal/gmol CO_2) + (-22,063 cal/gmol HCl) + (-57,798 cal/gmol $H_2O(g)$) - (-20,630 cal/gmol CH_3Cl) = **-153,283 cal/gmol**

For the dichloromethane reaction:

CH_2Cl_2: $CH_2Cl_2 + O_2 \rightarrow CO_2 + 2\ HCl$

$\Delta Hc°$ = (-94,052 cal/gmol CO_2) + 2 (-22,063 cal/gmol HCl) - (-22,800 cal/gmol CH_2Cl_2) = **-115,378 cal/gmol**

For the chloroform reaction:

$CHCl_3$: $CHCl_3 + 7/6\ O_2 \rightarrow CO_2 + 1/3\ H_2O + 1/3\ HCl + 4/3\ Cl_2$

$\Delta Hc°$ = (-94,052 cal/gmol CO_2) + 1/3 (-57,798 cal/gmol $H_2O(g)$) + 1/3 (-22,063 cal/gmol HCl) - (-24,200 cal/gmol $CHCl_3$) = **-96,472 cal/gmol**

For the carbon tetrachloride reaction:

CCl_4: $CCl_4 + O_2 \rightarrow CO_2 + 2\ Cl_2$

$\Delta Hc°$ = (-94,052 cal/gmol CO_2) - 2 (24,000 cal/gmol CCl_4) = **-46,052 cal/gmol**

These data can be plotted to show the change in the heat of combustion of this homologous series of chlorinated compounds as a function of the number of chlorine atoms in the molecule as indicated below:

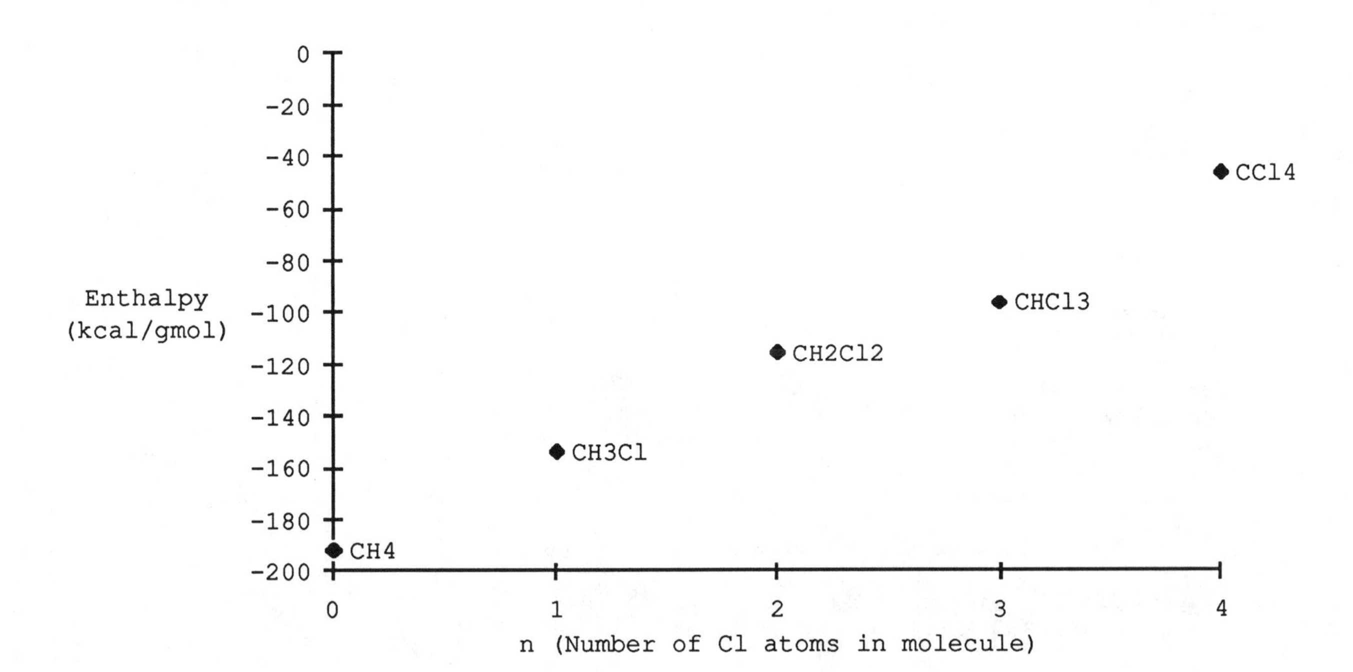

T-13 Solution.

The air requirements are calculated for the diesel fuel and #2 fuel oil for 1 hour of operation. For this basis, there is 10^4 lb of contaminated soil.

Mass of diesel fuel on soil = 0.025 wt% (10,000 lb soil) = 250 lb diesel fuel

Mass of #2 fuel oil = 20 lb

The composition of the combined fuel is calculated based on compositional mass fraction data given in the problem statement using the following calculations:

$$n = \left(\frac{20 \text{ lb fuel oil}}{250 \text{ lb diesel}}\right) = 0.08$$

$$C = \frac{C_{diesel} + n\,(C_{fuel\ oil})}{1 + n} = \frac{0.872 + 0.08\ (0.866)}{1 + 0.08} = 0.8716 \text{ lb C/lb feed}$$

$$H = 0.1214 \text{ lb H/lb feed}$$
$$S = 0.0069 \text{ lb S/lb feed}$$

The stoichiometric oxygen requirement is then calculated from the EPA worksheets:

Oxygen requirement for C:
$C\ (2.67 \text{ lb } O_2/\text{lb C}) = (2.67 \text{ lb } O_2/\text{lb C})\ (0.8716 \text{ lb C/lb feed})$
$= 2.327 \text{ lb } O_2/\text{lb feed}$
Oxygen requirement for H:
$H\ (8 \text{ lb } O_2/\text{lb H}) = (8 \text{ lb } O_2/\text{lb H})\ (0.1214 \text{ lb H/lb feed})$
$= 0.9712 \text{ lb } O_2/\text{lb feed}$
Oxygen requirement for S:
$S\ (1.0 \text{ lb } O_2/\text{lb S}) = (1.0 \text{ lb } O_2/\text{lb S})\ (0.0069 \text{ lb S/lb feed})$
$= 0.0069 \text{ lb } O_2/\text{lb feed}$

Total oxygen requirement = 2.327 + 0.9712 + 0.0069 = **3.31 lb O_2/lb feed**

The combustion gas mass flow rate is now calculated:

$$CO_2 = C\left(3.67\ \frac{\text{lb } CO_2}{\text{lb C}}\right) = 0.8716\ \frac{\text{lb C}}{\text{lb feed}}\left(3.67\ \frac{\text{lb } CO_2}{\text{lb C}}\right) = 3.2 \text{ lb } CO_2/\text{lb feed}$$

$$H_2O = H\ (9 \text{ lb } H_2O/\text{lb H}) = 1.093 \text{ lb } H_2O/\text{lb feed}$$

$$SO_2 = S\ (2 \text{ lb } H_2O/\text{lb H}) = 0.0138 \text{ lb } SO_2/\text{lb feed}$$

$$N_2 = O_{2\ stoichiometric}\left(3.31\ \frac{\text{lb } N_2}{\text{lb } O_2}\right) = 3.31\ \frac{\text{lb } O_2}{\text{lb feed}}\left(3.31\ \frac{\text{lb } N_2}{\text{lb } O_2}\right)$$

$$= 10.94 \text{ lb } N_2/\text{lb feed}$$

Total combustion product gas mass flow rate = 3.2 + 1.093 + 0.0138 + 10.94 = **15.25 lb gas/lb feed**

The mass fraction of each component in the combustion gas is calculated below:

T-13 Solution (Continued).

$$CO_2 = \frac{3.2 \text{ lb } CO_2/\text{ lb feed}}{15.25 \text{ lb gas/lb feed}} = 0.2098$$

$$H_2O = 0.0717,\ SO_2 = 9.0 \times 10^{-4},\ N_2 = 0.7174$$

NOTE: Since the SO_2 is present at < 1% in the combustion gas, it has been neglected in further calculations.

The volume of each component in the flue gas is calculated next.

NOTE: This calculation is not required to answer the problem at hand, but is the next step in the EPA worksheet method. It is also of relevance to calculations involving the design of pollution control equipment.

$$Q_{CO_2} = \text{Mass fraction of } CO_2 \text{ in combustion gas} \left(\frac{\text{Mass of combustion gas}}{\text{Density of } CO_2}\right)$$

$$= 0.2098 \left(\frac{15.25 \text{ lb combustion gas/lb feed}}{0.114 \text{ lb } CO_2/\text{scf}}\right) = 28.07 \text{ scf } CO_2/\text{lb feed}$$

$$Q_{H_2O} = 0.0717 \left(\frac{15.25 \text{ lb combustion gas/lb feed}}{0.0467 \text{ lb } H_2O/\text{scf}}\right) = 23.41 \text{ scf } H_2O/\text{lb feed}$$

$$Q_{N_2} = 0.7174 \left(\frac{15.25 \text{ lb combustion gas/lb feed}}{0.0727 \text{ lb } N_2/\text{scf}}\right) = 150.49 \text{ scf } N_2/\text{lb feed}$$

Total combustion gas flow rate = 28.07 + 23.41 + 150.49 = **202.0 scf/lb feed**

Rigorous thermochemical calculations are then carried out to determine the final operating temperature of the incinerator. Assume that the air, waste and fuel are all initially at 60°F.

Cooling Step

$$\Delta H_{air} = 0,\ \Delta H_{waste} = 0,\ \Delta H_{fuel} = 0$$

Combustion Step

The mass fraction of each component is calculated based on a total mass of 10,000 lb soil/diesel + 20 lb fuel oil.

$$m_{soil} = 9750/10{,}020 = 0.9732$$
$$m_{diesel} = 250/10{,}020 = 0.02495$$
$$m_{fuel\ oil} = 20/10{,}020 = 0.002$$

The NHV is calculated for the mixture of materials entering the incinerator based on the weighted sum of the NHVs of the three components above.

$$NHV_T = m_{soil}\ (NHV_{soil}) + m_{diesel}\ (NHV_{diesel}) + m_{fuel\ oil}\ (NHV_{fuel\ oil})$$
$$= 0.9732\ (0) + 0.02495\ (18{,}600 \text{ Btu/lb}) + 0.002\ (19{,}500 \text{ Btu/lb})$$
$$= \mathbf{503.1\ Btu/lb}$$

T-13 Solution (Continued).

Heating Step

The ΔH values for the product gases are calculated. For this analysis, the heat capacities are assumed to be independent of temperature. A more rigorous solution would include this dependence.

$$\Delta H_p = \sum_{products} m_i c_{Pi} (T - T_o)$$

The mass fraction of each component leaving the incinerator is calculated for the product gas ΔH calculation.

Total mass = $M_{soil} + M_{CO_2} + M_{H_2O} + M_{N_2} + M_{SO_2}$ = 9750 lb soil + (3.2 lb CO_2/lb feed + 1.093 lb H_2O/lb feed + 10.94 lb N_2/lb feed) (250 + 20) lb feed

Total mass = **13,863 lb product**

The mass fraction of each component then is:

m_{soil} = 9750/13,863 = 0.7033
m_{CO_2} = 864/13,863 = 0.0623
m_{H_2O} = 295.1/13,863 = 0.0213
m_{N_2} = 2953.8/13,863 = 0.2131

The enthalpy change with this combustion product gas composition is:

$$\Delta H_P = \left(0.7033 \frac{\text{lb soil}}{\text{lb product}} \left(0.30 \frac{\text{Btu}}{\text{lb-°F}}\right) + 0.0623 \frac{\text{lb } CO_2}{\text{lb product}} \left(0.27 \frac{\text{Btu}}{\text{lb-°F}}\right) \right.$$

$$\left. + 0.0213 \frac{\text{lb } H_2O}{\text{lb product}} \left(0.52 \frac{\text{Btu}}{\text{lb-°F}}\right) + 0.2708 \frac{\text{lb } N_2}{\text{lb product}} \left(0.27 \frac{\text{Btu}}{\text{lb-°F}}\right)\right)(T - T_o)$$

ΔH_P is equated to NHV_T, and the resulting equation is solved for T. The equation simplifies to:

$$0.312 (T - T_o) = 503.1 \text{ Btu/lb}$$

yielding a final combustion temperature = **1672 °F**

T-14 Solution.

First write the relevant chemical equilibrium reaction:

$$\frac{1}{2} N_{2(g)} + \frac{1}{2} O_{2(g)} \leftrightarrow NO_{(g)} \tag{1}$$

From the equilibrium constant described below:

T-14 Solution (Continued).

$$K = \frac{P_{NO}}{P_{N_2}^{1/2} P_{O_2}^{1/2}} = \frac{P_{NO} P_{N_2}^{1/2}}{P_{N_2} P_{O_2}^{1/2}} \tag{2}$$

The following expression is obtained by taking the log of both sides of Equation 2:

$$\log \frac{P_{NO}}{P_{N_2}} = \log K + \frac{1}{2} \log \frac{P_{O_2}}{P_{N_2}} \tag{3}$$

Equation 3 represents the distribution of N_2, O_2 and NO as a function of K and the ratio of P_{O_2}/P_{N_2}. Assuming that this ratio is constant, which is valid at all except the highest temperatures, log P_{O_2}/P_{N_2} then depends only on the temperature dependence of K.

As a first approximation, the temperature dependence of K can be determined by using the van't Hoff equation.

The equilibrium constant K and its temperature dependence can be determined by knowing $\Delta G°$, $\Delta H°$, and Δa, Δb, and Δc. (Values can be obtained from Appendixes I, H, and F, respectively).

$\Delta G° = 20.69$ kcal/gmol
$\Delta H° = 21.60$ kcal/gmol
$\Delta a = 0.035$
$\Delta b = -3.0 \times 10^{-5}$
$\Delta c = 1.2 \times 10^{4}$

The equilibrium constant at 298 K can be determined by:

$$\log K_{298} = \frac{-\Delta G°}{2.3\ RT_{298}} = -15.21$$

The temperature dependence of K can be determined from:

$$\frac{d\ \ln(K)}{dT} = \frac{\Delta H°}{RT^2} \tag{4}$$

If $\Delta H°$ is assumed to be independent of temperature, Equation 4 can be integrated to yield:

$$\ln \frac{K_1}{K_2} = -\frac{\Delta H°}{R}\left(\frac{1}{T_1} - \frac{1}{T_2}\right) \tag{5}$$

Conversion to log base 10, and substitution into Equation 3 yields the following expression:

T-14 Solution (Continued).

$$\log\left(\frac{P_{NO}}{P_{N_2}}\right) = \log K_{298} - \frac{\Delta H}{2.3\ R}\left(\frac{1}{T} - \frac{1}{298}\right) + \frac{1}{2}\log\left(\frac{P_{O_2}}{P_{N_2}}\right) \quad (6)$$

Assuming a constant ratio of P_{O_2}/P_{N_2}, and collecting terms, Equation 6 becomes:

$$\log\left(\frac{P_{NO}}{P_{N_2}}\right) = \left(\frac{\Delta H° - \Delta G°}{2.3\ R\ 298}\right) + \frac{1}{2}\log\left(\frac{0.21}{0.79}\right) - \frac{\Delta H}{2.3\ R\ T} \quad (7)$$

A more accurate calculation can be made which allows $\Delta H°$ to vary with temperature by using the definition of heat capacity:

$$dH = C_p dT \quad (8)$$

The temperature dependence of C_p can be expressed as:

$$Cp = a + bT + cT^{-2} \quad (9)$$

Then for the overall chemical reaction, after substituting Equation 9 into Equation 8 and integrating:

$$\Delta H = \Delta H° + \Delta a\ T + \frac{\Delta b}{2}T^2 - \frac{\Delta c}{T} \quad (10)$$

When Equation 10 is substituted into Equation 4 and integrated, the following expression results:

$$\ln\frac{K_1}{K_2} = \frac{-\Delta H°}{R}\left(\frac{1}{T_1} - \frac{1}{T_2}\right) + \frac{\Delta a}{R}\ln\left(\frac{T_1}{T_2}\right) + \frac{\Delta b}{2R}(T_1 - T_2) + \frac{\Delta c}{2R}\left(\frac{1}{T_1^2} - \frac{1}{T_2^2}\right) \quad (11)$$

When Equation 11 is converted to log base 10, terms are collected, and the results are substituted into Equation 3, the following equation results:

$$\log\left(\frac{P_{NO}}{P_{N_2}}\right) = \frac{(\Delta H° - \Delta G°)}{2.3\ R\ (298)} + \frac{1}{2}\log\left(\frac{P_{O_2}}{P_{N_2}}\right) - \frac{\Delta H°}{2.3\ R\ T} + \frac{\Delta a \log T}{2.3\ R} + \frac{\Delta b\ T}{2\ (2.3)\ R}$$

$$+ \frac{\Delta c}{2\ (2.3)\ R\ T^2} + \left[-\frac{\Delta a\log(298)}{2.3\ R} - \frac{\Delta b\ (298)}{2\ (2.3)\ R} - \frac{\Delta c}{2\ (2.3)\ R\ (298)^2}\right] \quad (12)$$

The results for the solution of Equations 7 and 12 as a function of combustion temperature are presented in the table below, while Equation 7 results are shown graphically in the figure that follows. The calculations show that as temperature increases, the ratio of P_{O_2}/P_{N_2} also increases asymptotically to 1.0, at which point, $P_{O_2} = P_{N_2}$.

T-14 Solution (Continued).

Temperature (K)	$\log(P_{O_2}/P_{N_2})$ (Equation 7)	$\log(P_{O_2}/P_{N_2})$ (Equation 12)
298	-15.5	-15.5
500	-9.08	-9.09
1000	-4.35	-4.36
1500	-2.77	-2.79
2000	-1.98	-2.00
2500	-1.51	-1.53
3000	-1.2	-1.22

These results clearly show that the thermodynamic equilibrium shifts to the right in Reaction 1 as temperature increases. Note that this analysis neglects the kinetics of formation and decomposition of Reaction 1 which are quite complicated.

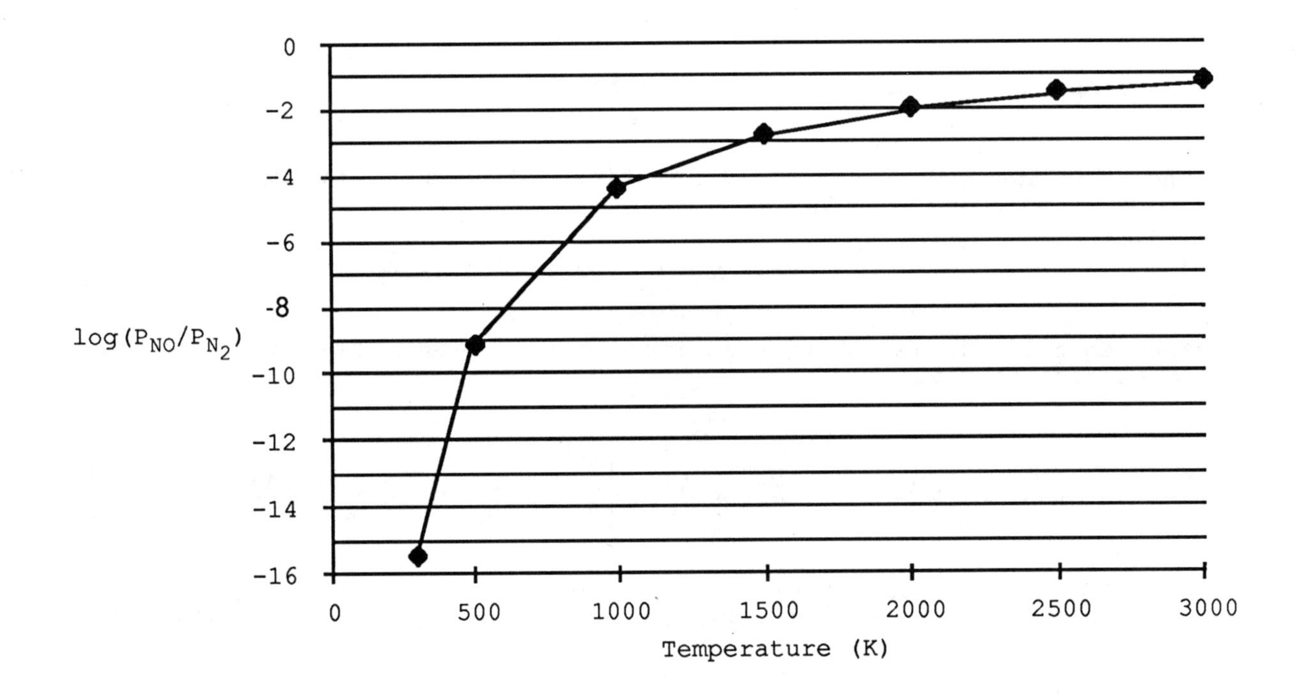

T-15 Solution.

From the graph of $\log(P_{O_2}/P_{N_2})$ versus temperature, it follows that at temperatures of 1500 to 2000 °C it does not matter in which form the nitrogen initially exists. If the nitrogen was converted to N_2, it will be in equilibrium with NO; otherwise it will likely be converted to NO, which would then equilibrate with N_2. Either way, the NO concentration is a function of temperature and not feedstock when air is used for combustion.

T-16 Solution.

Rearranging Equation 1 (see problem statement) and solving for the desired rate constant k yields:

$k = -\ln(1-\eta)/t = -\ln(1 - 0.995)/0.5 =$ **10.6 1/s**

Rearranging Equation 2 for the calculation of T yields:

$$T = \frac{-E}{R\left(\ln\left(\frac{k}{A}\right)\right)} \qquad (5)$$

Toluene Calculations

Calculating E from Equation 4 yields:

$E = -0.00966$ (92 g/gmol) + 46.1 = 45.2 kcal/gmol = **45,200 cal/gmol**

From Equation 3 assuming 1 atm pressure and a mole fraction of oxygen of 0.15 yields a value of A of:

$$A = \frac{1 \text{ atm } (0.15)\ (2.85 \times 10^{11})}{0.08205 \text{ L-atm/gmol-K}}\left(\frac{16}{92 \text{ g/gmol}}\right) = 9.06 \times 10^{10}$$

$k/A = 10.6/(9.06 \times 10^{11}) = 1.17 \times 10^{-10}$

$\ln(k/A) = -22.89$

Substituting into Equation 5 yields:

$T = -45{,}200$ cal/gmol/[(1.987 cal/gmol-K)(-22.89)] = **994 K = 1330 °F**

Benzene Calculations

Calculating E from Equation 4 yields:

$E = -0.00966$ (78 g/gmol) + 46.1 = 45.35 kcal/gmol = **45,350 cal/gmol**

From Equation 3 assuming 1 atm pressure and a mole fraction of oxygen of 0.15 yields a value of A of:

$$A = \frac{1 \text{ atm } (0.15)\ (2.85 \times 10^{11})}{0.08205 \text{ L-atm/gmol-K}}\left(\frac{16}{78 \text{ g/gmol}}\right) = 9.0 \times 10^{10}$$

$k/A = 10.6/(9.0 \times 10^{11}) = 1.18 \times 10^{-10}$

$\ln(k/A) = -22.86$

Substituting into Euqation 5 yields:

$T = -45{,}350$ cal/gmol/[(1.987 cal/gmol-K)(-22.86)] = **998.3 K = 1337 °F**

T-17 Solution.

Integration of the kinetic expression in compartment i gives:

$$\ln\left(\frac{C_i}{C_{i-1}}\right) = -k\ (t_i - d) \tag{1}$$

where C_i is the concentration leaving the ith hearth. When Equation 1 is summed over n chambers in the multiple hearth system, the following equation results:

$$\ln\left(\frac{C_n}{C_o}\right) = nkd - k\sum_{i=1}^{n} t_i \tag{2}$$

a. The required retention time is calculated based on rearrangement of Equation 2 as indicated below:

$$t_r = \sum_{i=1}^{n} t_i = nd - \frac{\ln(C_n/C_o)}{k} \tag{3}$$

The reaction rate constant, k, is calculated as follows:

$k = 3 \times 10^7/s \exp[(-40{,}000\ cal/gmol)/((1.9872\ cal/gmol\text{-}K)(1255.22\ K))]$
$= \mathbf{3.26/s}$

The the total required reaction time is determined from Equation 3 as:

$t_r = 4\ (2.5\ s - 0.001\ s/°F\ (1800\ °F)) - (1/3.26\ s)\ (\ln\ (0.0001))$
$= 4\ (0.7\ s) - (-2.83\ s) = \mathbf{5.62\ s}$

b. As in Part a above, the retention time can be expressed as:

$$\ln\left(\frac{C_n}{C_o}\right) = -\sum_{i=1}^{n} k_i t_i + \sum_{i=1}^{n} k_i d_i \tag{4}$$

Since the reaction times in each hearth are equal, $t_i = t_r/n$. Then,

$$\ln\left(\frac{C_n}{C_o}\right) = -(t_r/n)\sum_{i=1}^{n} k_i + \sum_{i=1}^{n} k_i d_i \tag{5}$$

Substitution yields:

$\ln\ (0.0001) = -\ (t_r/2)\ (3/s + 5/s) + (3/s\ (1\ s) + 5/s\ (3\ s))$
$-9.21 = -\ (t_r/2)\ (8/s) + (18);$
$\mathbf{t_r = 6.8\ s}$

T-18 Solution.

In this problem, the reaction rate varies along the combustion chamber due to the temperature gradient which occurs. The reaction rate constant, k, is calculated as follows:

$k = 2 \times 10^8$/s exp[(-50,000 cal/gmol)/((1.9872 cal/gmol-K) (1366.3 K))]
= **2.01/s**

Length	Temperature (K)	k (1/s)
0	1366.3	2.01
10	1355.2	1.73
20	1344.1	1.48
30	1333.0	1.27
40	1321.9	1.08
50	1310.8	0.92
60	1299.7	0.78
70	1288.6	0.66
80	1277.4	0.56
90	1266.3	0.47
100	1255.2	0.39

As a first approximation it can be assumed that k = 1.0/s. The first order kinetic expression integrated over the total retention time of the incinerator is:

$$\frac{dC}{dt} = -k\,C(t)\ ;\ \ln\left(\frac{C_{t_r}}{C_o}\right) = -kt_r \tag{1}$$

Solving for t_r with C_{t_r}/C_o = 0.0001 yields:

ln(0.0001) = - 1.0/s (t_r); -9.21 = - 1.0/s (t_r); t_r = 9.21 s

Therefore, the nominal length = 9.21 s (10 ft/s) = **92.1 ft.**

For a rigorous solution, the kinetic expression is written as:

$$\frac{dC}{dt} = -A\exp\left(\frac{-E}{R\,(T_o - avt)}\right) C(t) \tag{2}$$

where T_o is the temperature at the flame, a is the temperature gradient, and v is the gas velocity. Integration of this equation yields:

$$\ln\left(\frac{C_{t_r}}{C_o}\right) = -\int_0^{t_r} k(t)\,dt \tag{3}$$

where k(t) = A exp(-E/(R (T_o - avt))). From the trapezoidal rule:

$$\int_C^D k(t)\,dt = 0.5\,(D - C)\,(k(C) + k(D)) \tag{4}$$

T-18 Solution (Continued).

Substituting Equation 4 into Equation 3 and solving for t_r yields:

$$\ln\left(\frac{C_{t_r}}{C_o}\right) = -0.5A\,(t_r - 0)\left(\exp\left(\frac{-E}{R\,(1366.3 - 1.11(0))}\right) + \exp\left(\frac{-E}{R\,(1366.3 - 1.11t_r)}\right)\right)$$

$$\ln(0.0001) = -0.5(2 \times 10^8/\text{s})\,(t_r)\left(\exp\left(\frac{-50{,}000 \text{ cal/gmol}}{1.9872 \text{ cal/gmol-K }(1366.3 \text{ K})}\right) + \exp\left(\frac{-50{,}000 \text{ cal/gmol}}{1.9872 \text{ cal/gmol-K }(1366.3 \text{ K} - 1.11\,t_r)}\right)\right)$$

$$-9.2103 = -1 \times 10^8/\text{s}\,(t_r)\left(\exp(-18.42) + \exp\left(\frac{-25{,}161}{(1366.3 - 1.11\,t_r)}\right)\right)$$

$$t_r \exp\left(\frac{-25{,}161}{(1366.3 - 1.11\,t_r)}\right) + t_r\,(1.0 \times 10^{-8}) - 9.2103 \times 10^{-8} = 0$$

Solving this equation by trial and error yields a value of t_r = 9.2103 s for 99.99% destruction, with a minimal incinerator length of **92.10 ft**.

CHAPTER FIVE

Incinerators

I-1 Solution.

a. Neglect the 0.1% hexachlorobenzene for calculation purposes. The stoichiometry for the combustion of Wastes A and B are as follows:

Waste A:

$$C_{30}H_{60} + n\ (O_2 + 3.76\ N_2) \rightarrow 30\ CO_2 + 30\ H_2O + 3.76n\ N_2$$

where n = 45 lbmol/lbmol A

Waste B:

$$C_{15}H_{15}Cl_5O_3 + m\ (O_2 + 3.76\ N_2) \rightarrow 15\ CO_2 + 5\ H_2O + 5\ HCl + 3.76\ m\ N_2$$

where m = 16 lbmol/lbmol B

The feed mixture based on a 100 lbmol basis = F = 20 A + 80 B

The oxygen requirement is then calculated as:

20 (45) + 80 (16) (0.9) = 2052 lbmol oxygen

For 50% EA, oxygen requirement = **3078 lbmol oxygen**

The air supplied to the system = 3078 (4.76) = **14,652 lbmol air**

For the new feed mixture based on a 100 lbmol basis = F = 40 A + 60 B

The oxygen requirement is then calculated as:

40 (45) + 60 (16) (0.9) = 2664 lbmol oxygen

For 50% EA, oxygen requirement = **3996 lbmol oxygen**

Thus, the air supplied for the 20%A case is adequate (3078 versus 2664 lbmol), but the EA is reduced to 16% (3078/2664) if the air rate is not changed.

b. In order to estimate the effect of changing feed composition on incinerator temperature (without a change in air flow rate), the net heating value and combustion temperatures must be calculated. NHV can be calculated from DuLong's equation as indicated below:

$$\text{NHV (Btu/lb)} = 14{,}000\ (m_C) + 45{,}000\ (m_H - 0.125\ m_O) - 760\ (m_{Cl})$$

With the 20% A composition:

m_C = 0.2 (360/420) + 0.8 (180/420) = 0.514
m_H = 0.2 (60/420) + 0.8 (15/420) = 0.058
m_O = 0.2 (0) + 0.8 (48/420) = 0.091
m_{Cl} = 0.2 (0) + 0.8 (177.5/420) = 0.338

With the 40% A composition:

m_C = 0.4 (360/420) + 0.6 (180/420) = 0.60
m_H = 0.4 (60/420) + 0.6 (15/420) = 0.078
m_O = 0.4 (0) + 0.6 (48/420) = 0.069
m_{Cl} = 0.4 (0) + 0.6 (177.5/420) = 0.254

I-1 (Continued).

With the 20% A composition:

NHV = 14,000 (0.514) + 45,000 (0.058 - 0.125 (0.091)) - 760 (0.338)
= 7196 + 2070 - 257 = **9009 Btu/lb**

With the 40% A composition:

NHV = 14,000 (0.60) + 45,000 (0.078 - 0.125 (0.069)) - 760 (0.254)
= 8400 + 3121 - 193 = **11,328 Btu/lb**

Combustion temperature can be estimated using the following equation:

$$T = 60 + \frac{NHV}{[0.3\ \{1 + (1 + EA)\ (7.5 \times 10^{-4})\ (NHV)\}]}$$

With the 20% A composition:

$$T = 60 + \frac{9009}{[0.3\ \{1 + (1 + 0.5)(7.5 \times 10^{-4})(9009)\}]} = \mathbf{2756\ °F}$$

With the 40% A composition:

$$T = 60 + \frac{11,328}{[0.3\ \{1 + (1 + 0.16)\ (7.5 \times 10^{-4})\ (11,328)\}]} = \mathbf{3538\ °F}$$

Thus, the temperature increase is very high (can damage the incinerator) and since the EA is minimal, incomplete combustion may also result.

I-2 Solution.

a. Determine the constituent mass fractions of the waste feed in the absence of water in order to calculate the NHV of the waste using DuLong's equation. With 100 lb of the granular activated carbon as a basis, the following mass and mass fractions can be calculated:

	Mass Fraction
m_C = 100 lb + 31 lb/100 lb (72/163) = 113.7 lb C	113.7/137 = 0.83
m_{ash} = 6 lb ash	6/137 = 0.044
m_H = 31 lb/100 lb (4/163) = 0.761 lb H	0.761/137 = 0.006
m_O = 31 lb/100 lb (16/163) = 3.04 lb O	3.04/137 = 0.022
m_{Cl} = 31 lb/100 lb (71/163) = 13.5 lb Cl	13.5/137 = 0.099

From the DuLong's equation:

NHV = 14,000 (m_C) + 45,000 (m_H - 0.125 m_O) - 760 (m_{Cl})

NHV = 14,000 (0.83) + 45,000 [0.006 - 0.125 (0.022)] - 760 (0.099)
= **11,700 Btu/lb dry basis**

NHV = [137/(137+10)] (11,700 Btu/lb dry) - (10/147) (1050 Btu/lb H_2O) {heat of vaporization} = **10,833 Btu/lb wet basis**

With 5% heat loss from the system,
the final NHV = 0.95 (10,833) = **10,291 Btu/lb wet basis**

I-2 (Continued).

Using a mean specific heat value of 0.325 Btu/lb-°R, the required waste NHV to produce an incinerator temperature of 2400 °F is estimated as follows:

$$T = 60 + \frac{NHV}{0.325\ \{1 + (1 + EA)(7.5 \times 10^{-4})(NHV)\}}$$

$$2400 = 60 + \frac{NHV\ (1 - 0.05)}{0.325\{1 + (1 + 0.5)(7.54 \times 10^{-4})(NHV)(1 - 0.05)\}}$$

NHV = **5723 Btu/lb dry**

The amount of water to cool the waste to 2400 °F is based on the addition of X lb of water to the original wet carbon to reduce the NHV from 10,833 Btu/lb to 5723 Btu/lb, i.e.,

$$5723\ \text{Btu/lb} = 10{,}833\ \text{Btu/lb}\left(\frac{137}{147 + X}\right) - 1050\ \text{Btu/lb}\ H_2O\left(\frac{10 + X}{147 + X}\right)$$

X = 93.4 lb water/lb wet carbon = 11.2 gal water/lb wet carbon to provide the required NHV for a 2400 °F combustion temperature.

b. The pounds per hour of wet carbon plus water (NHV = 5723 Btu/lb) for a heat release rate of 25,000 Btu/h-ft^3 is calculated based on the dimensions of the incinerator as:

Volume of incinerator = π (6 ft)^2/4 (18 ft) = 508.9 ft^3
Total heat release = 25,000 Btu/h-ft^3 (508.9 ft^3) = 12,723,450 Btu/h
Water + carbon loading rate = 12,723,450 Btu/h/5723 Btu/lb
= 2223 lb water + carbon/h
Original carbon loading rate = (147/(147 + 93.4)) 2223
= 1359 lb wet carbon/h

c. Other approaches for cooling must either reduce the waste NHV, or increase the EA. These methods might include:

- Increasing the EA with more combustion air
- Blend in other low NHV material (i.e., contaminated soil, contaminated water) for treatment as well as cooling

Water has a high heat capacity and a much smaller volume of it is required for cooling than would EA. EA is easy to move and might be a simple option, not requiring an additional feed system. The use of contaminated material of low heat capacity provides treatment as well as cooling, but requires adequate blending and a constant and reliable source of this material. The most appropriate solution may be to provide some water cooling, options for high EA inputs, and contingencies for feeding additional low NHV wastes to reduce water and EA requirements.

I-3 Solution.

Heat transfer by conduction is given by:

$$q = \frac{\Delta T}{R}$$

where q = heat transfer rate (Btu/h)
ΔT = temperature differential (°F)
R = heat transfer resistance term (°F-h/Btu).

For cylindrical geometry, R is defined as:

$$R = \frac{\ln (r_o/r_i)}{2 \pi L k}$$

where r_o = outside radius of cylinder (ft)
r_i = inside radius of cylinder (ft)
L = length of cylinder (ft)
k = thermal conductivity (Btu/h-°F-ft)

The q for the ordinary brick is first calculated using the thickness of the fire brick assumed to be = 0.5 ft. The expression for the ordinary brick q is:

$$q = \frac{(700 - 100\ °F)\ (2\ \pi\ L)\ (0.82\ Btu/h\text{-}ft\text{-}°F)}{\ln \left(\frac{10.83\ ft}{10.5\ ft} \right)} \tag{1}$$

q = **99,913.6 L Btu/h**

The q for the fire brick is then calculated using the thickness of the fire brick, X:

$$q = \frac{(3000 - 700\ °F)\ (2\pi L)\ (0.028\ Btu/h\text{-}ft\text{-}°F)}{\ln \left(\frac{10 + X}{10} \right)} \tag{2}$$

$$q = \frac{404.637\ L}{\ln \left(\frac{10 + X}{10} \right)}$$

Equating the results of Equation 2 to Equation 1 and solving for X yields a fire brick thickness of **X = 0.041 ft**

Since this value of X is not equivalent to the assumed value of X, the calculations should be repeated with the new value of X. The ordinary brick q is found to be:

$$q = \frac{(700 - 100\ °F)\ (2\ \pi\ L)\ (0.82\ Btu/h\text{-}ft\text{-}°F)}{\ln \left(\frac{10.37\ ft}{10.04\ ft} \right)}$$

= **95,588.3 L Btu/h**

I-3 (Continued).

Equating this result to Equation 3 and solving for X yields a new calculated fire brick thickness of **X = 0.042 ft**

Since the new value is essentially equal to the assumed value, the thickness of the fire brick is **0.042 ft = 0.5 in.**

I-4 Solution.

a. In a plug flow model, the residence time for a first order, irreversible reaction can be described as:

$$t = -\frac{1}{k} \ln \left(\frac{C_{At}}{C_{Ao}} \right)$$

For 99.995% destruction, $C_{At}/C_{Ao} = 5 \times 10^{-5}$, so that t becomes:

$$t = -\frac{1}{38.3/s} \ln (0.00005) = 0.259 \text{ s}$$

Reactor volume = 12.1 L/s (0.259 s) = **3.13 L plug flow model**

b. In a continuously stirred tank reactor model, the kinetic expression is:

$$t = \frac{C_{At} - C_{Ao}}{-k\, C_{At}} = \frac{\frac{C_{At}}{C_{Ao}} - 1}{-k \left(\frac{C_{At}}{C_{Ao}} \right)}$$

For 99.995% destruction, $C_{At}/C_{Ao} = 5 \times 10^{-5}$, so that t becomes:

$$t = \frac{0.00005 - 1}{-38.3/s\ (0.00005)} = 519 \text{ s}$$

Reactor volume = 12.1 L/s (519 s) = **6286 L CSTR model**

In order to reduce the volume of the CSTR required, two CSTRs in series can be utilized as illustrated below:

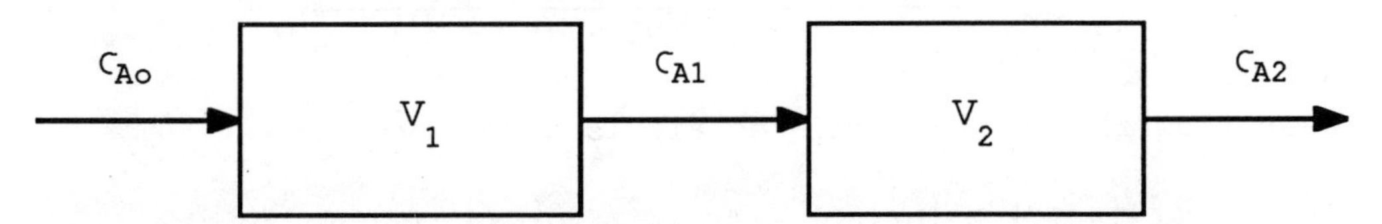

Mass balances for Reactors 1 and 2 in series can be written as:

$$0 = C_{Ao} - C_{A1} - k\, C_{A1}\, t_1 \qquad (1)$$

$$0 = C_{A1} - C_{A2} - k\, C_{A2}\, t_2 \qquad (2)$$

I-4 (Continued).

From Equation 1,

$$C_{A1}/C_{Ao} = 1/(1 + k\ t_1) \qquad (3)$$

From Equation 2,

$$C_{A2}/C_{A1} = 1/(1 + k\ t_2) \qquad (4)$$

Assuming $t_1 = t_2$, and multiplying Equations 3 and 4 to eliminate C_{A1} yields:

$$C_{A2}/C_{Ao} = 1/(1 + k\ t_2)^2 \qquad (5)$$

For 99,995% destruction, $C_{At}/C_{Ao} = 5 \times 10^{-5}$, so that t becomes:

$$0.00005 = 1/(1 + 38.3/s\ t)^2 = 3.6\ s$$

The total reactor volume required under this configuration = 2 (12.1 L/s) (3.6 s) = **87 L for the CSTR 2 reactor model**

I-5 Solution.

a. A mass balance may be written as:

$$\dot{m}_{PA} + \dot{m}_m = \dot{m}_E \qquad (1)$$

where $\dot{m}_{PA}$ = mass flow rate of polluted air (lb/min)
$\dot{m}_m$ = mass flow rate of methane (lb/min)
$\dot{m}_E$ = mass flow rate of exhaust (lb/min)

An energy balance is given by:

$$\dot{m}_A h_A + \dot{m}_m h_m + \dot{m}_m\ (1 - f_L)\ NHV_m + \dot{m}_P\ (1 - f_L)\ NHV_P = \dot{m}_E h_E \qquad (2)$$

where $\dot{m}_P$ = mass flow rate of component P in the air stream (lb/min)
h = enthalpy of the PA, m and E (Btu/lbmol)
f_L = fractional heat loss in incinerator
NHV= net heating value of m and P (Btu/lb)

Solving Equations 1 and 2 simultaneously and rearranging yields the following equation:

$$\dot{m}_m = \frac{\dot{m}_{PA}\ (h_E - h_{PA}) - \dot{m}_P\ (1 - f_L)\ NHV_P}{(1 - f_L)\ NHV_m - (h_E - h_m)} \qquad (3)$$

The mass flow rate of the polluted air is given by:

$$\dot{m}_{PA} = \rho_{PA}\ Q = \frac{(14.7 + 1\ lb/in^2)\ (144\ in^2/ft^2)\ (40{,}000\ ft^3/min)}{\dfrac{1545\ ft\text{-}lb/lbmol\text{-}°R}{29\ lb/lbmol}\ (125 + 460\ °R)} = 2902\ lb/min$$

The exhaust gas and methane enthalpies are essentially the same as air, and flue gas typically has a molecular weight of 30.

I-5 (Continued).

From Appendix D, enthalpy values are determined as follows:

$$h_E = (14{,}970 \text{ Btu/lbmol})/(30 \text{ lb/lbmol}) = 499 \text{ Btu/lb}$$

The polluted air at 125 °F requires interpolation of Appendix D values as shown below:

$$h_{PA} = \frac{472.7 \text{ Btu/lbmol} + \dfrac{1170 - 472.7 \text{ Btu/lbmol}}{4}}{29 \text{ lb/lbmol}} = 22.3 \text{ Btu/lb}$$

From Appendix C, at 70 °F and a reference temperature = 32 °F,

$$8.31 \frac{\text{Btu}}{\text{lbmol-°R}} \frac{1 \text{ lbmol}}{16 \text{ lb}} (70 - 32 \text{ °R}) = 20.8 \text{ Btu/lb}$$

For 100 lbmol/min mass flow of the pollutant, the mass flow rate is:

$$\dot{m}_P = (100 \text{ lbmol/min})\ (30 \text{ parts}/10^6 \text{ parts})\ (112.5 \text{ lb chlorobenzene/lbmol})$$
$$= 0.3375 \text{ lb/min}$$

By DuLong's equation:

$$NHV = 14{,}000\ (m_C) + 45{,}000\ (m_H - 0.125\ m_O) - 760\ (m_{Cl}) = \text{Btu/lb}$$

$$NHV_P = 14{,}000\ (0.64) + 45{,}000\ (0.04444) - 760\ (0.3155) = 10{,}720 \text{ Btu/lb}$$

From Appendix G, NHV_m = 21,520 Btu/lb. Substituting these data into Equation 3 yields:

$$\dot{m}_m = \frac{2902 \text{ lb/min } (499 - 20.8 \text{ Btu/lb}) - 0.3375 \text{ lb/min } (1 - 0.95)\ (10{,}720 \text{ Btu/lb})}{(1 - 0.05)\ (21{,}520 \text{ Btu/lb}) - (499 \text{ Btu/lb} - 499 \text{ Btu/lb})}$$

$$= 67.9 \text{ lb/min}$$

$$\rho_m = \frac{(100 + 14.7 \text{ lb/in}^2)\ (144 \text{ in}^2/\text{ft}^2)}{\dfrac{1545 \text{ ft-lb/lbmol-°R}}{16 \text{ lbmol/lb}} (70 + 460 \text{ °R})} = 0.323 \text{ lb/ft}^3 \text{ @ 70 °F and 100 psig}$$

$$Q_m = (67.9 \text{ lb/min}/0.323 \text{ lb/ft}^3) = \mathbf{210 \text{ cfm @ 70 °F and 100 psig}}$$

b. Gas density in the HFI with an approximate molecular weight of 30 is:

$$\rho = \frac{14.7 \text{ lb/in}^2\ (144 \text{ in}^2/\text{ft}^2)}{\dfrac{1545 \text{ ft-lb/lbmol-°R}}{30 \text{ lb/lbmol}} (2000 + 460 \text{ °R})} = 0.0167 \text{ lb/ft}^3$$

The volumetric flow rate through the HFI is:

$$Q_T = \frac{\Sigma \dot{M}}{\rho} = \frac{2902 + 67.9 + 0.3375 \text{ lb/min}}{0.0167 \text{ lb/ft}^3} = 1.78 \times 10^5 \text{ cfm}$$

I-5 (Continued).

Combining $L/D = 3$, $ut_r = L$ and $Q_T = (\pi u D^2)/4$ yields:

$$D = \left(\frac{4\ t_r\ Q}{3\pi}\right)^{0.333} = \left(\frac{4\ (1.4\ s)\ (1.78 \times 10^5\ cfm)\ (1\ min/60\ s)}{3\pi}\right)^{0.333} = \mathbf{12\ ft}$$

c. The heat release rate in Btu/h-ft^3:

$$= \frac{[0.3375\ lb/min\ (10{,}720\ Btu/lb) + 69.3\ lb/min\ (21{,}520\ Btu/lb)]\ 60\ min/h}{\frac{\pi\ (12)^2}{4}\ (3)\ (12\ ft)}$$

$= \mathbf{21,600\ Btu/h\text{-}ft^3}$

I-6 Solution.

a. Determine the mass fractions of the waste feed components using a basis of 100 lb clean #6 fuel oil:

$mass_C = 86.7 + 1\ (84/161) = 87.222$ lb/101 lb mixture; $m_C = 0.8636$
$mass_H = 10.3 + 1\ (6/161) = 10.337$ lb; $m_H = 0.1023$
$m_O = 0$
$mass_{Cl} = 0 + 1\ (71/161) = 0.441$ lb; $m_{Cl} = 0.0044$
$mass_S = 3.0 + 0 = 3.0$ lb; $m_S = 0.0297$

The NHV is estimated using DuLong's formula:

$$NHV = 14{,}000\ (m_C) + 45{,}000\ (m_H - 0.125\ m_O) - 760\ (m_{Cl}) + 4500\ (m_S)$$

$$NHV = 14{,}000\ (0.8636) + 45{,}000\ [0.1023 - 0.125\ (0)] - 760\ (0.0044) + 4500\ (0.0297) = 16{,}824\ Btu/lb$$

The operating temperature for the incinerator operated at 25% EA is calculated using the following equation:

$$T = 60 + \frac{NHV}{[0.325\{1 + (1 + EA)\ (7.5 \times 10^{-4})\ (NHV)\}]}$$

$$T = 60 + \frac{16{,}824\ (1 - 0.05)}{[0.325\{1 + (1 + 0.25)(7.54 \times 10^{-4})(16{,}824)(1 - 0.05)\}]} = \mathbf{3121\ °F}$$

b. The EA calculation is based on the above equation, with a given T = 3121 - 300 = 2821°F. This solution is as follows:

$$2821 = 60 + \frac{16{,}824\ (1 - 0.05)}{[0.325\ \{1 + (1 + EA)\ (7.54 \times 10^{-4})\ (16{,}824)(1 - 0.05)\}]}$$

EA = **39.5%**

I-6 (Continued).

c. The heat release rate is used to calculate L and D as follows:

$$\frac{25{,}000\ \text{Btu/h/ft}^3}{16{,}284\ \text{Btu/lb}}\ \frac{\pi\ D^2\ \text{ft}^2}{4}\ (3\ D\ \text{ft}) = 25{,}000\ \text{lb/h}$$

$D = 8.94$ ft ∴ **$L = 26.82$ ft**

I-7 Solution.

a. The volume of solids is calculated from their mass and density as:

Volume = 500 kg/1500 kg/m^3 = 0.333 m^3 solids

With $e_1 = 0$, L_1 is determined from the solids volume as

$L_1 = 0.333\ \text{m}^3/0.5\text{m}^2 = 0.666$ m

Then with $e_{mf} = e_2 = 0.56$, L_{mf} is calculated using the equation from Geankoplis as:

$0.666\ \text{m}/L_{mf} = (1 - 0.56)/(1 - 0)$; L_{mf} = **1.51 m**

b. The physical properties of air at 800 K and 1.5 atm are:

$\mu = 4.3 \times 10^{-5}$ Pa-s
$\rho = 0.66$ kg/m^3
$P = 1.52 \times 10^5$ Pa

The pressure drop at minimum fluidizing conditions from Geankoplis is than calculated as:

$\Delta P = 1.51\ \text{m}\ (1 - 0.56)(1500 - 0.66\ \text{kg/m}^3)\ 9.81\ \text{m/s}^2 = 0.0977 \times 10^5$ Pa
= **0.096 atm**

c. The Reynolds Number is calculated at the minimum fluidization velocity using the equation by Wen and Yu:

$$N_{Re} = \left[33.7^2 + 0.0408\ (0.00015\ \text{m})^3\ \frac{0.66\ \text{kg/m}^3\ (1500 - 0.66\ \text{kg/m}^3)\ 9.81\ \text{m/s}^2}{(4.3 \times 10^{-5}\ \text{Pa-s})^2}\right]^{1/2} - 33.7$$

N_{Re} = **0.0107**

Through the definition of the Reynolds Number, the minimum fluidization velocity may be determined as:

$N_{Re} = D_p(v_{mf})\rho/\mu = 0.00015\ \text{m}\ v_{mf}\ (0.66\ \text{kg/m}^3)/4.3 \times 10^{-5}\ \text{Pa-s} = 0.0107$

$v_{mf} = 0.0107\ (4.3 \times 10^{-5}\ \text{Pa-s})/(0.00015\ \text{m}\ (0.66\ \text{kg/m}^3)) = 0.0047$ m/s
= **0.0154 fps**

I-8 Solution.

a. The combustion of cellulose waste has the following reaction stoichiometry:

$$C_6H_{10}O_5 + 6\ O_2 \rightarrow 6\ CO_2 + 5\ H_2O$$
$$CaO + H_2O \rightarrow Ca(OH)_2$$

Mass Balance Calculations:

Water in feed limestone: 1 mol water/mol $Ca(OH)_2$ produced
[(18 lb/lbmol water)/(74.1 lb/lbmol $Ca(OH)_2$)] (40 lb/h)
= 9.73 lb water/h

Total dry feed: 800 lb/h - 9.73 lb water/h = 790.27 lb dry feed/h

Ash(CaO): 1 mol CaO/mol $Ca(OH)_2$ used
[(56.06 lb/lbmol CaO)/(74.1 lb/lbmol $Ca(OH)_2$)] (40 lb/h)
= 30.27 lb CaO/h

Cellulose: 760 lb/h

Stoichiometric air:
[(760 lb cellulose/h)/(162.1 lb/lbmol cellulose)] (6 mol O_2/lbmol cellulose) (32 lb/lbmol O_2) (4.32 lb air/lb O_2) = 3887 lb air/h
Total air: at 100% EA, 3887 lb/h (1 + 1) = 7774 lb air/h

Water in combustion air: at 5%, 7774 lb air/h (0.005) = 39 lb water/h

Water produced in combustion reaction:
[(760 lb cellulose/h)/(162.1 lb/lbmol cellulose)] (5 lbmol H_2O/lbmol cellulose) (18 lb/lbmol H_2O) = 422 lb H_2O/h

Total water in flue gas: 9.73 + 39 + 422 lb H_2O/h = **471 lb H_2O/h**

Carbon dioxide in flue gas:
[(760 lb cellulose/h)/(162.1 lb/lbmol cellulose)] (6 lbmol CO_2/lbmol cellulose) (44 lb/lbmol CO_2) = **1238 lb CO_2/h**

Nitrogen in flue gas: Total air mass - O_2 mass
[7774 lb air/h - (760 lb cellulose/h)/(162.1 lb/lbmol cellulose)] (12 lbmol $O_{2\ @\ 100\%EA}$/lbmol cellulose) (32 lb/lbmol O_2) = **5974 lb N_2/h**

Oxygen in flue gas:
[(760 lb cellulose/h)/(162.1 lb/lbmol cellulose)] (6 lbmol O_2/lbmol cellulose) (32 lb/lbmol O_2) = **900 lb O_2/h**

b. Using Dulong's approximation, the heating value of cellulose can be found to be 6250 Btu/lb cellulose. The heating value of the limestone is found to be:

Σ(heats of formation) = $H_{f_{H_2O}} + H_{f_{CaO}} - H_{f_{Ca(OH)_2}}$
= (-57.8 kcal/gmol) + (-151.8 kcal/gmol) - (-239.7 kcal/gmol)
= 30.1 kcal/gmol = 730 Btu/lb lime

Latent heat of vaporization of moisture associated with lime:
970 Btu/lb water vaporized (18 lb/lbmol water/74.1 lb/lbmol $Ca(OH)_2$)
= 235 Btu/lb lime

I-8 (Continued).

Thus, the net heating value of the cellulose/lime mixture is determined as follows:

NHV = 6250 Btu/lb (760 lb cellulose/800 lb total) - (730 + 235 Btu/lb lime) (40 lb cellulose/800 lb total) = **5890 Btu/lb mixture**

Using the NHV relationship given in the problem statement, the estimated temperature of the combustion system is found as:

$$T = 60 + \frac{NHV}{[0.325 \{1 + (1 + EA)(7.5 \times 10^{-4})(NHV)\}]}$$

$$T = 60 + \frac{5890}{[0.325 \{1 + (1 + 1)(7.5 \times 10^{-4})(5890)\}]} = \mathbf{1903\ ^\circ F}$$

c. The required EA for a bed temperature (assumed uniform) of 2200 °F is calculated using the following form of the NHV equation given in the problem statement:

$$EA = \frac{\left[\frac{NHV}{0.3\ (T - 60)}\right] - 1}{7.5 \times 10^{-4}\ (NHV)} - 1 = EA = \frac{\left[\frac{5890}{0.3\,(2200 - 60)}\right] - 1}{7.5 \times 10^{-4}\ (5890)} - 1 = 0.85 = 85\%\ EA$$

d. From inspection of the equation above for temperature in Part b, it can be found that as the NHV tends to infinity, temperature approaches an upper limit of 2282 °F. Thus, no finite amount of fuel can raise the combustion temperature to 2300 °F.

I-9 Solution.

Assume the waste enters the top of the kiln and then falls to the bottom. This takes place in one-half of a revolution, for a total of 2/N minutes.

If the solid falls vertically during this 2/N minutes, it will have travelled ("horizontally") a total distance of $\Delta L = D \sin\alpha \approx D\ \alpha$ (for small α). See diagram below:

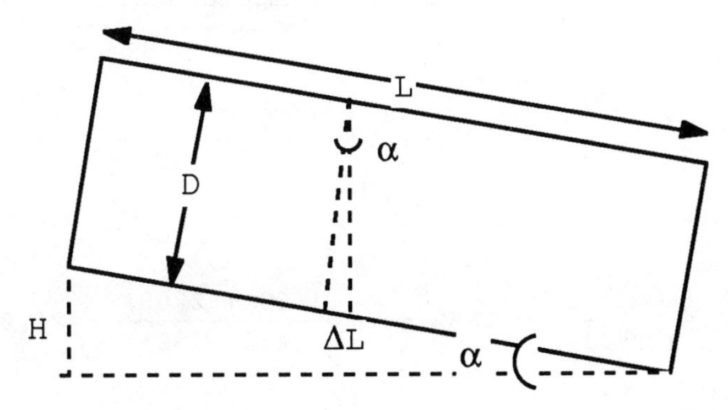

Slope, $S = H/L \approx \alpha$

I-9 (Continued).

The linear velocity of the solid inside the kiln is V, fpm, and is expressed as:

$$V = \frac{D\,\alpha}{2/N} = \frac{DSN}{2}$$

The residence time, or the time required to travel L ft can be described as:

$$\theta = \frac{L}{\frac{DSN}{2}} = \frac{2\,L}{DSN}$$

This can be compared to the following expression based on experimental data:

$$\theta = \frac{0.19\,L}{DSN}$$

I-10 Solution.

As the percentage of PCB increases in the waste mixture, the following will occur:

1. Incinerator temperature will decrease because of the low heat of combustion of PCBs.
2. The efficiency of combustion will decrease because the combustion temperature decreases, and PCBs are more difficult to incinerate at a lower temperature.
3. The concentration of Cl_2 in the flue gas will increase because at lower temperatures, the HCl $\Leftrightarrow$ Cl_2 equilibrium is shifted toward Cl_2.

I-11 Solution.

a. From the Ideal Gas Law, PQ = RT.
   Q = 350 scf/lb (1000 lb/h) (1 h/60 min) = 5833 scfm
   PQ/T = R, therefore Q/T = R/P = constant

Hence:

$Q_{60}/T_{60} = Q_T/T$, where T is absolute temperature.

@ 1700 °F

$$Q_{1700} = Q_{60}\frac{T}{T_{60}} = 5833\text{ scfm}\,\frac{(460 + 1700)}{(460 + 60)} = \mathbf{24{,}230\ acfm}$$

@ 2200 °F

$$Q_{2200} = Q_{60}\frac{T}{T_{60}} = 5833\text{ scfm}\,\frac{(460 + 2200)}{(460 + 60)} = \mathbf{29{,}840\ acfm}$$

b. Temperature adjusted rate constants from laboratory data are determined as follows:

I-11 (Continued).

@ 1700 °F

$$k = A\,e^{-\frac{E_{act}}{RT}} = 3.3 \times 10^{7}/s\; e^{\frac{-25{,}000\ cal/gmol}{1.982\ cal/gmol\text{-}L\ (T)}} = \mathbf{899/s}$$

@ 2200°F

$$k = \mathbf{6451/s}$$

c. $V = L/\theta = Q/A$, hence $L = V\,\theta = 20$ fps (2 s) = **40 ft.**

Then $A = Q\theta/L = \pi D^2/4$; solving for diameter, D,

$$D = \left[\frac{4\ Q\ \theta}{L\ \pi}\right]^{0.5}$$

@ 1700 °F

$$D = \left[\frac{4\ (24{,}230\ acfm)\ (2\ s)\ (1\ min/60\ s)}{\pi\ (40\ ft)}\right]^{0.5} = \mathbf{5.07\ ft}$$

@ 2200 °F

$$D = \mathbf{5.63\ ft}$$

d. For a continuous flow stirred tank reactor:

$$\frac{C}{C_o} = (1 + k\theta)^{-1}$$

Hence the removal efficiency is:

$$\frac{C_o - C}{C_o} = 1 - \frac{C}{C_o} = 1 - [1 + k\ \theta]^{-1}$$

$efficiency_{1700}$ = **99.994%** (4 - 9's)

$efficiency_{2200}$ = **99.9992%** (5 - 9's)

e. For a plug flow reactor (i.e., no longitudinal mixing)

$$\ln\left(\frac{C}{C_o}\right) = -k\theta$$

Converting to common logs yields,

$$\log \frac{C}{C_o} = -\frac{k\ \theta}{2.303}$$

@ 1700 °F

$$\left(\frac{C}{C_o}\right)_{1700} = 10^{-781} \Rightarrow \quad (781\ 9's)$$

I-11 (Continued).

@ 2200 °F

$$\left(\frac{C}{C_o}\right)_{2200} = 10^{-5602} \Rightarrow (5602\ 9\text{'s})$$

NOTE: The plug flow reactor is predicted to provide much better removal; however, there is so much turbulence in this type of system, that this level of improved performance will not be observed in practice.

I-12 Solution.

A sludge loading rate of 6 T/h and 10% EA are used for the calculation of the amount of methane required for the drying and recalcination process.

Evaporative Water:

[6 T/h (2000 lb/T) (0.5 lb water/lb sludge) (1059.9 Btu/lb water + 9.17 Btu/lbmol-°F {Cp @ 1800 °F}) (1 lbmol water/18 lb) (1800 - 60 °F)]
= 1.17×10^7 Btu/h

Calcining Reaction:

$$CaCO_3 \rightarrow CaO + CO_2$$

From Appendix H, the ΔH value for calcination is found from heat of formation calculations:

ΔH_f = (-152 kcal/mol CaO) + (-94 kcal/mol CO_2) - (-288 kcal/mol $CaCO_3$)
= 42 kcal/gmol = 79,000 Btu/lbmol

Since 25% of the sludge is $CaCO_3$, the calcining reaction will require:

6 T/h (2000 lb/T) (0.25 lb $CaCO_3$/lb sludge) (1 lbmol/100 lb $CaCO_3$) (79,000 Btu/lbmol) = 2.37×10^6 Btu/h

Sensible heat required for the ash (ash = inerts in feed + 1 mole CaO/mole $CaCO_3$):

6 T/h (2000 lb/T) (0.25 lb ash/lb sludge) + (30 lbmol CaO/h) (56 lb CaO/lbmol CaO) = 4680 lb ash/h

Heat required to raise ash from 60 °F to 1800 °F:

4680 lb ash/h (0.22 Btu/lb-°F) (1800 - 60 °F) = 1.79×10^6 Btu/h

To find the methane required, the heat released from the combustion of the methane is equated to the heat required for evaporation, calcining plus all sensible heat requirements.

From Appendix G, for methane, stoichiometric air requirements are 17.27 lb air/lb methane, the NHV is 21,520 Btu/lb CH_4 and the products are found to be:

$$2.74\ \text{lb}\ CO_2/\text{lb}\ CH_4 + 2.25\ \text{lb}\ H_2O/\text{lb}\ CH_4 + 13.28\ \text{lb}\ N_2/\text{lb}\ CH_4$$

I-12 (Continued).

At 10% EA, 19 lb air/lb methane are required, with the combustion products being:

2.74 lb CO_2/lb CH_4 + 2.25 lb H_2O/lb CH_4 + 1.73 lb O_2/lb CH_4 + 14.6 lb N_2/lb CH_4

For solution of this energy balance, let Y = lb methane burned. For each Y lb methane combusted, Y (21,520 Btu) are generated.

The heat required, assuming constant C_ps, is: $Y (\Sigma m C_p) (1800 - 60)$

Cps @ 1800 °F are:

CO_2	11.75 Btu/lbmol-°F/44 lb/lbmol = 0.27 Btu/lb-°F
H_2O	9.17 Btu/lbmol-°F/18 lb/lbmol = 0.51 Btu/lb-°F
N_2	7.46 Btu/lbmol-°F/28 lb/lbmol = 0.27 Btu/lb-°F
O_2	7.88 Btu/lbmol-°F/32 lb/lbmol = 0.25 Btu/lb-°F

The sensible heat of the flue products is then calculated as:

Y [(2.74 lb CO_2/lb CH_4) (0.247 Btu/lb CO_2-°F) + (2.25 lb H_2O/lb CH_4) (0.51 Btu/lb H_2O-°F) + (14.6 lb N_2/lb CH_4) (0.27 Btu/lb N_2-°F) + (1.73 lb O_2/lb CH_4) (0.25 Btu/lb O_2-°F)] (1800 - 60 °F) = Y (1.09×10^4 Btu/lb CH_4)

Equating energy production to energy requirements, assuming no loss in the kiln yields:

Y (21,520 Btu/lb CH_4) = 1.17×10^7 Btu/h + 2.37×10^6 Btu/h + 1.79×10^6 Btu/h + Y (1.09×10^4 Btu/lb CH_4)

(1.06×10^4) Y Btu/lb CH_4 = 1.59×10^7 Btu/h; **Y = 1500 lb CH_4/h**

CHAPTER SIX

Waste Heat Boilers/Quenchers

WHB/Q-1 Solution.

a. The size of the heat exchanger must first be calculated. The heat transfer area in ft^2 is calculated using the equation given in the problem statement

$$q = w_h C_p (T_{H1} - T_{H2}) = U\ A\ \Delta T_{ln\ mean}$$

$\Delta T_{ln\ mean}$ is calculated using the gas temperature given in the problem statement:

$$\Delta T_{ln\ mean} = \frac{(2000 - 380\ °F) - (380 - 328\ °F)}{\ln\left(\dfrac{2000 - 328\ °F}{380 - 328\ °F}\right)} = \mathbf{466.8\ °F}$$

The heat capacity of the gas is based on a weighted average of the heat capacities of its constituents.

C_p = 0.8 (7.2 Btu/lbmol N_2-°F) + 0.1 (7.62 Btu/lbmol O_2-°F) + 0.1 (11.02 Btu /lbmol CO_2-°F) = **7.62 Btu/lbmol-°F**

The average molecular weight of the gas mixture is also based on a weighted average of its constituents and is:

MW = 0.8 (28 lb/lbmol N_2) + 0.1 (32 lb/lbmol O_2) + 0.1 (44 lb/lbmol CO_2)
= **30 lb/lbmol air**

Therefore, the mass basis C_p = (7.62 Btu/lbmol air-°F)/30 lb/lbmol
= 0.254 Btu/lb-°F

Boiler tube area = π D L N = π (1 in)/(12 in/ft) 20 ft N = 5.236 ft^2 N

Substituting the results from above into the expression for overall heat transfer yields the following expression:

50,000 lb/h (0.254 Btu/lb-°F) (2000 - 380 °F)
= $(1986/N)^{0.8}$ (5.236 ft^2 N) (466.8 °F)

Solving for N yields the following:

$N^{0.2}$ = [50,000 (0.265) (2000 - 380)]/[1986 (5.236) (466.8) = **3.20**
N = $(3.20)^5$ = **336 tubes**

b. This problem is solved using the nomograph found in Appendix M. First, the temperature difference ratio, ϕ, is calculated:

$\phi = (T_{H1} - T_{H2})/(T_{H2} - T_c)$ = (2000 - 380 °F)/(380 - 328 °F) = **31.2**

Then the average flue gas temperature T_{Ha} is calculated:

T_{Ha} = (2000 + 380 °F)/2 = **1190 °F**

From Appendix M, w_h/N = **144 lb/h-tube**

Solving for N from this result yields: N = w_h/144
= (50,000 lb/h)/144 lb/h-tube
= **347 tubes which is fair agreement with Part a.**

WHB/Q-2 Solution.

a. The size of the heat exchanger must first be calculated. The heat transfer area in ft^2 is calculated using the following equation:

$$q = U_o A_o \Delta T_m \quad (1)$$

where q = heat transfer rate (Btu/h)
U_o = overall heat transfer coefficient (Btu/h-ft^2-°F)
A_o = total heat transfer area (ft^2)
ΔT_m = average temperature driving force (°F)

Assuming a counter-current flow, $\Delta T_m = \Delta T_{ln\ mean}$ (lnmean temperature difference, °F), and is defined as:

$$\Delta T_m = \frac{\Delta T_1 - \Delta T_2}{\left(\ln \dfrac{\Delta T_1}{\Delta T_2} \right)} \quad (2)$$

where ΔT_1 = temperature differential between the inlet and outlet hot gas temperature = $T_{h\ inlet} - T_{h\ outlet}$ (°F)
ΔT_2 = temperature differential between the inlet and outlet cold gas temperature = $T_{c\ inlet} - T_{c\ outlet}$ (°F)

The hot gas inlet temperature = 400 °F, the hot gas outlet temperature must be calculated, the cold gas inlet temperature = 120 °F, and the cold gas outlet temperature = 220 °F.

Using these data, Equation 1 may also be written as:

$$q = m_1 C_{p1} (T_{h\ inlet} - T_{h\ outlet}) = m_1 C_{p1} (T_{c\ inlet} - T_{c\ outlet})$$

where m_1 = mass flow rate of hot gas = ρV = 0.05 lb/ft^3 (10,000 acfm) = **500 lb/min**
m_2 = mass flow rate of cold gas = m_1 + water added

water added = absolute humidity of cold gas - absolute humidity of hot gas
= 0.082 lb H_2O/lb dry gas (from Appendix A) - 0.04 lb H_2O/lb dry gas = **0.042 lb H_2O/lb dry gas**

Therefore m_2 = 500 lb/min + 0.042 lb H_2O/lb dry gas (500 lb/min)
= **521 lb/min**

The hot gas stream inlet temperature is then calculated from the heat transfer equation:

$$(500\ \text{lb/min})\ (0.25\ \text{Btu/lb-°F})\ (T_{h\ inlet} - 400\ \text{°F}) = (521\ \text{lb/min})\ (0.30\ \text{Btu/lb-°F})\ (120 - 400\ \text{°F})$$

$T_{h\ inlet}$ = 400 - 125 = **275 °F**

The heat transfer area is then calculated by rearrangement of Equation 1 as:

WHB/Q-2 Solution (Continued).

$$A_o = \frac{q}{U_o \, \Delta T_m}$$

ΔT_m is calculated using the hot gas outlet temperature:

$$\Delta T_m = \frac{(400 - 220) - (275 - 120)}{\ln\left(\dfrac{400 - 220}{275 - 120}\right)} = \mathbf{167\ ^\circ F}$$

The the overall heat transfer rate is:

q = 500 lb/min (0.25 Btu/lb-°F) (400 - 275 °F)
= 15,625 Btu/min = 938,000 Btu/h

Then the required heat transfer area is:

A = 938,000 Btu/h)/[(25 Btu/h-ft^2-°F) (167 °F)] = 224.7 = **225 ft^2**

The heat exchanger costs can then be calculated using the equation given in the problem statement as:

$$C \approx 40{,}000\ (225\ ft^2)^{-0.44}\ e^{0.067\ [\ln(225\ ft^2)]^2} = \mathbf{\$26{,}000}$$

with the installed cost ≈ 3 (C) = **$78,000**

b. The additional required fan power is estimated by the additional work required to overcome the additional head loss due to the installation of the heat exchanger, i.e.,

W = V ΔP = 10,000 acfm (0.2 psi) (144 in^2/ft^2) = **288,000 ft-lb/min**

Power required = 288,000 ft-lb/min (60 min/h) (0.746 kw/33,000 ft-lb/min)
= **390 kwh (100% fan efficiency) = 651 kwh (60% fan efficiency)**

The required power cost is then = 651 kwh ($0.10/kwh)
= **$65.10/h = $1560/d = $546,000/yr**

WHB/Q-3 Solution.

a. Heat loss during cooling of the combustion gases (q_1) goes to sensible and latent heat of the quench water (q_2), i.e., $q_1 = q_2$. The heat loss to cool the gases is calculated as follows:

$$q_1 = \Sigma C_p\ \Delta T_{\text{of mixture to be cooled}}$$

Using a basis of 100 lbmol, the energy loss due to combustion gas cooling in the quencher is:

q_1 = (100 lbmol) (8.67 Btu/lbmol-°F) (2000 - T °F) = 867 (2000 - T)

q_2 = w [(latent heat of vaporization of water @ 70°F) + ($h_{yT} - h_{y70}$)]
= w [(1054.3 Btu/lb) + ($h_{yT} - h_{y70}$)]

WHB/Q-3 Solution (Continued).

where w = lb water added to combustion gas following cooling
h_{yT} = enthalpy at temperature T (Btu/lb)
h_{yT} = enthalpy at temperature 70 °F = 1092.3 Btu/lb from Appendix F

q_2 = w (1054.3 Btu/lb + h_{yT} - 1092.3 Btu/lb) = w (h_{yT} - 38 Btu/lb)

A trial and error procedure is used to solve this energy balance problem. An initial guess of the final temperature, T, of 180 °F is made yielding the following results:

The saturation vapor pressure of water in the combustion gas at 180 °F is 7.51 psi from steam tables, which represents (7.51/14.7) (100) = 51.1 mol% water.

The mol% water also can be equated to:

$$\text{mol\% water} = \left(\frac{X}{70 + X}\right) 100$$

where X = final mol water = [(100 - 30 mol dry gas) + X] (% water/100)

Rearranging this equation yields an expression for X:

$$X = \frac{70\ (\text{mol\% water})}{100 - \text{mol\% water}} = \frac{70\ (51.1)}{100 - 51.1} = 73.15 \text{ mol}$$

The lb water added to the gas phase = (73.15 - 30 mol) (18 lb/lbmol) = **776.7 lb**

From enthalpy tables, h_{y180} = 1138.1 Btu/lb

Therefore, the energy balance can be solved using the data calculated above:

$$q_1 = 867\ (2000 - 180\ ^\circ\text{F}) = 1{,}577{,}940 \text{ Btu}$$

and

$$q_2 = 776.7 \text{ lb } (1138.1 - 38 \text{ Btu/lb}) = 854{,}434 \text{ Btu}$$

These heat values do not balance, and another guess must be made, and the calculations must be repeated until the results converge.

A second guess of 190 °F is made. Here, h_{y190} = 1442 Btu/lb, and the saturation vapor pressure = **9.339 psi,** or (9.3391/14.7) (100) = **63.5 mol% water. X is found to be 122 moles,** with a total of **1653 lb water added to the gas phase.**

Once again, the energy balance can be solved for using the revised data:

$$q_1 = 867\ (2000 - 190\ ^\circ\text{F}) = 1{,}569{,}270 \text{ Btu}$$

and

$$q_2 = 1653 \text{ lb } (1142 - 38 \text{ Btu/lb}) = 1{,}824{,}912 \text{ Btu}$$

WHB/Q-3 Solution (Continued).

An interpolation can be made between the two guesses to yield an approximate final temperature of **187°F**, yielding a **saturation vapor pressure of 8.79 psi.**

b. The final gas with the saturation vapor pressure **8.79 psi** yields a mol% (8.79/14.7) (100) = **59.8 mol% water. X is found to be 104.1 moles,** with a total of **1334 lb water added per 100 mol gas passed through the quencher.**

c. The volumetric flow rate of the exit gases is calculated based on the total combustion gas and evaporated quench water flow rates:

$$\text{Total gas to quencher} = 40{,}000 \text{ acfm} \left(\frac{460 + 32}{460 + 2000}\right)\left(\frac{1 \text{ lbmol}}{359 \text{ ft}^3}\right) = \mathbf{22.28 \text{ lbmol/min}}$$

$$\text{Water added} = 22.28 \text{ lbmol/min } (1334 \text{ lb water/100 lbmol gas}) = \mathbf{297 \text{ lb/min}}$$

Volumetric flow rate of exit gas is:

$$= \left(22.28 \text{ lbmol/min} + \frac{297 \text{ lb water/min}}{18 \text{ lb/lbmol water}}\right)\left(\frac{359 \text{ ft}^3}{\text{lbmol}}\right)\left(\frac{460 + 187}{460 + 32}\right) = \mathbf{18{,}315 \text{ acfm}}$$

WHB/Q-4 Solution.

A mass balance is written for each portion of the waste heat boiler system using a 1 h basis for all calculations.

The feed mass balance is written as:

$$F = HS + C + MW$$

where F = feed (lb/h)
HS = mass flow rate of steam to heat feed (lb/h)
C = mass flow rate of steam condensate (lb/h)
MW = mass flow rate of make-up feed water (lb/h)

The boiler mass balance is written as:

$$F = BD + S$$

where BD = mass flow rate of blow-down (lb/h)
S = mass flow rate of steam to turbine (lb/h)
MW = mass flow rate of make-up feed water (lb/h)

The feed energy balance is written as:

$$(HS)\ (h_s) + (C)\ (h_c) = (MW)\ (h_{MW}) + (F)\ (h_f)$$

where h = enthalpy of stream identified by subscript (Btu/lb)

The boiler energy balance is written as:

Q = available heat after radiation loss

$$Q = (S)\ (h_s) + (BD)\ (h_{BD}) - (F)\ (h_f)$$

WHB/Q-4 Solution (Continued).

From data presented in the problem statement, the following relationships can be written:

$$HS = 0.1\ S \qquad BD = (1/20)\ F$$

From steam tables and Appendix D, the following enthalpy and water saturation values can be found at the conditions given in the problem statement:

h_s = 1198 Btu/lb, h_c = 148 Btu/lb, h_f = 185 Btu/lb, h_{MW} = 36 Btu/lb
h_{BD} = 357 Btu/lb
h_{air}(dry 1800 °F) = 442 Btu/lb, $h_{moisture}$(1800 °F) = 2223 Btu/lb
h_{air} (dry 522 °F) at outlet = 111 Btu/lb, $h_{moisture}$ (552 °F) = 1372 Btu/lb

Saturated steam temperature @ 200 psia = 382 °F

With this final temperature, the outlet gas temperature is calculated as:

$$\text{Outlet gas temperature} = (382 + 140\ °F) = 522\ °F$$

The total heat balance for the exit gases is calculated based on the change in enthalpy of the cooled gas plus the change in enthalpy associated with the moisture added to the flue gas stream, i.e.,

$$q_{Total} = \text{mass gas flow rate } (h_{in} - h_{out}) + \text{mass moisture flow rate } (h_{in} - h_{out})$$

$$q_{Total} = 10{,}000 \text{ lb gas/h } (442 - 111 \text{ Btu/lb}) + 1500 \text{ lb water/h } (2223 - 1372 \text{ Btu/lb}) = 4.586 \times 10^6 \text{ Btu/h}$$

Radiation losses from the system (1%) are used to reduce the total energy of the system as follows:

$$q_{net} = = 0.99\ (4.586 \times 10^6 \text{ Btu/h}) = 4.541 \times 10^6 \text{ Btu/h}$$

Returning to the mass balance relationships, the following can be generated:

$$F = BD + S,\ BD = (1/20)\ F, \text{ therefore } S = (19/20)\ F \text{ and } F = (20/19)S$$

$$BD = (1/20)\ F = 1/20\ (19/20)\ S = S/19$$

From the boiler energy balance:

$$Q = (S)\ (h_s) + (BD)\ (h_{BD}) - (F)\ (h_f) = (S)\ (h_s) + (S/19)\ h_{BD} + (20/19)\ (S)\ (h_f)$$

Solving for S yields:

$$S = \frac{q_{net}}{h_s + \frac{h_{BD}}{19} - \frac{20\ h_f}{19}} = \frac{4.541 \times 10^6 \text{ Btu/h}}{1198 \text{ Btu/lb} + \frac{357 \text{ Btu/lb}}{19} - \frac{20\ (185 \text{ Btu/lb})}{19}} = \mathbf{4443\ \frac{Btu}{h}}$$

Then,

$BD = S/19 =$ **234 lb/h**

$F = S + BD = 4443 + 234 =$ **4677 lb/h**

$HS = 0.1\ S =$ **444 lb/h**

$C + MS = F - HS = 4677 - 444 =$ **4233 lb/h**

WHB/Q-4 Solution (Continued).

Then from the feed energy balance,

$$(HS)\ (h_s) + (C)\ (h_c) = (MW)\ (h_{MW}) + (F)\ (h_f)$$

Substitution for C = 4233 - MW and for the respective enthalpies, rearrangement and simplification yields the following results:

MW = [4677 lb/h (185 Btu/lb) - 444 lb/h (1198 Btu/lb) - 4233 lb/h (198 Btu/lb)]/(36 - 148 Btu/lb) = **2617 lb/h**

and finally, C = 4233 lb/h - 2317 lb/h = **1616 lb/h**

CHAPTER SEVEN

Air Pollution Control Equipment

APCE-1 Solution.

Advantages and disadvantages of electrostatic precipitators

Advantages:

- Extremely high particulate (coarse and fine) collection efficiencies can be attained with relatively low expenditure of energy.
- Dry collection and solids disposal.
- Low pressure drop (typically less than 0.5 inches of water).
- Designed for continuous operation with minimum maintenance requirements.
- Relatively low operating costs.
- Capable of operation under high pressure (to 150 psi) or vacuum conditions.
- Capable of operation at high temperatures (to 1300 °F).
- Relatively large gas flow rates can be effectively handled.

Disadvantages

- High capital costs.
- Very sensitive to fluctuations in gas stream conditions (flow, temperature, particulate and gas composition, and particulate loading).
- Certain particulates are difficult to collect due to extremely high or low resistivity characteristics.
- Relatively large space requirements for installation.
- Explosion hazard when treating combustible gases and/or collecting combustible particulates.
- Special precautions are required to safeguard personnel from high voltage equipment.
- Ozone is produced by the negatively charged discharge electrode during gas ionization.
- Relatively sophisticated maintenance personnel required.

Advantages and disadvantages of fabric filter systems

Advantages:

- Extremely high particulate (coarse to submicron) collection efficiencies can be attained.
- Dry collection and solids disposal.
- Relatively insensitive to gas stream fluctuations. Efficiency and pressure drop are relatively unaffected by large changes in inlet dust loadings for continually cleaned filters.
- Corrosion and rusting of components usually not a problem.
- There is no hazard of high voltage, simplifying maintenance and repair, and permitting the collection of flammable dust.
- Use of selected fibrous or granular filter aids (precoating) permits the high-efficiency collection of submicron smokes and gaseous contaminants.
- Filter collectors are available in a large number of configurations, resulting in a range of dimensions and inlet and outlet flange locations to suit a wide range of installation requirements.
- Relatively simple operation.

Disadvantages

- Temperatures much in excess of 550 °F require special refractory mineral or metallic fabrics that are still in the developmental stages and can be very expensive.
- Certain particulates may require fabric treatments to reduce dust seeping, or to assist in the removal of the collected dust.

APCE-1 Solution (Continued).

Fabric Filter Disadvantages (Continued)

- Concentrations of some dusts in the collector (≈ 50 g/m^3) may represent a fire or explosion hazard if a spark or flame is admitted by accident. Fabrics can burn if readily oxidizable dust is being collected.
- Relatively high maintenance requirements (bag replacements, etc.).
- Fabric life may be shortened at elevated temperatures and in the presence of acid or alkaline particulate or gas components.
- Hydroscopic materials, condensation of moisture, or tarry, adhesive components may cause crusty caking or plugging of the fabric and may require special additives.
- Replacement of fabric may require respiratory protection for maintenance personnel.
- Medium pressure-drop requirements, typically in the range of 4 to 10 inches of water.

Advantages and disadvantages of venturi scrubbers

Advantages:

- Provides both wet and dry particle removal.
- Provides quenching of flue gas temperature.
- Provides simultaneous gas and particle removal.
- Enhances the removal of condensible materials.
- Low capital costs.

Disadvantages

- Produces a liquid effluent that must be further processed.
- Cools the flue gas resulting in increased production of visible plume.
- Increases the water vapor concentration in the flue gas.

Advantages and disadvantages of wet inonization precipitator

Advantages:

- Provides high particulate collection efficiency for both wet and dry particle removal.
- Provides some quenching of flue gas temperature.
- Provides simultaneous gas and particle removal.
- Enhances the removal of condensible materials.

Disadvantages

- Produces a liquid effluent that must be further processed.
- Cools the flue gas resulting in increased production of visible plume.
- Increases the water vapor concentration in the flue gas.
- High capital costs.

APCE-1 Solution (Continued).

Advantages and disadvantages of spray dryers

Advantages:

- Provides high HCl collection efficiency and moderate SO_2 collection efficiency.
- Compatible with fabric filters for simultaneous collection of gases and particles.
- Enhances the removal of condensible materials.
- Low pressure drop.

Disadvantages

- Produces a liquid effluent that must be further processed.
- Must operate at a dry bulb temperature greater than the saturation temperature.
- Cools the flue gas resulting in increased production of visible plume.

Advantages and disadvantages of wet scrubbers

Advantages:

- Relatively small space requirements.
- Has the ability to collect gases as well as particulates, especially the "sticky" ones.
- Has the ability to handle high-temperature, high-humidity gas streams.
- Low capital costs if wastewater treatment system is not required.
- Ability to achieve high collection efficiencies for fine particulates, at however the expense of increased pressure drop in the system.

Disadvantages

- Produces a liquid effluent that must be further processed.
- Flue gas becomes saturated with water vapor.
- Cools the flue gas resulting in increased production of visible plume.
- Pressure drop and horsepower requirements may be high.
- Solids build-up at the wet dry interface may be a problem.
- System has relatively high maintenance costs.

APCE-2 Solution.

Determine the properties of the gas at 1700 °F and the density of the particles.

μ_g = 0.122 lb/h-ft
ρ_g = 0.0192 lb/ft^3
ρ_p = 2.9 (62.4 lb/ft^3) = 180.96 lb/ft^3

Determine the velocity of the gas at the cyclone's inlet.

$V_i = Q_g/(H\ W)$
W = 0.25 (D) = 0.75 ft
V_i = (4520 cfm)/[(1.5 ft) (0.75 ft)] = **4020 cfm**

The cut diameter is calculated using the equation presented in the problem statement as:

APCE-2 (Continued).

$$d_{p,50} = \left(\frac{9\ (0.112\ \text{lb/h-ft})\ (0.75\ \text{ft})\ (1\ \text{h}/60\ \text{min})}{2\ \pi\ (6.0)\ (4020\ \text{cfm})\ (180.96 - 0.0192\ \text{lb/ft}^3)} \right)^{0.5} = 5.2 \times 10^{-5}\ \text{ft}$$

$$= \mathbf{16.1\ \mu m}$$

For 5 μm particles the cyclone's collection efficiency is calculated as:

$E_{5\mu m} = 1.0/[1.0 + (16.1/5)^2] = 0.088 =$ **8.8%**

These calculations are carried out for the balance of the particles in the gas stream to yield the following overall efficiency:

Mean Diameter (μm)	E	Mass Fraction In	Mass Fraction Out
5	0.088	0.25	0.23
25	0.71	0.50	0.15
60	0.93	0.25	0.02

The overall removal = 1 - mass fraction out = 1 - 0.40 = 0.60 = 60%, therefore the concentration of particulates in the exit gas stream = **2 gr/ft³**

APCE-3 Solution.

For 90% removal of 0.5 μm particles:

The mean droplet diameter is:

$d_{mean} = 16{,}400/(272\ \text{ft/s}) + 1.45\ R^{1.5} = 60.29 + 1.45\ R^{1.5}$

For the removal efficiency calculation, A is calculated as follows:

$A = [1.325\ (0.5\ \mu m)^2\ (43.7\ \text{lb/ft}^3)\ (272\ \text{ft/s})]/[(9)\ (1.5 \times 10^{-5}\ \text{lb/ft-s})\ (60.29 + 1.45\ R^{1.5})]$

Removal efficiency can then be expressed as:

$E = 1 - \exp\ [(-0.2\ \text{ft}^3/1000\ \text{gal})\ (R)\ \{[1.325\ (0.5\ \mu m)^2\ (43.7\ \text{lb/ft}^3)\ (272\ \text{ft/s})]/[(9)\ (1.5 \times 10^{-5}\ \text{lb/ft-s})\ (60.29 + 1.45\ R^{1.5})]\}^{0.5}]$

By trial-and-error, it can be found that particle removal efficiency = 90% when **R = 14 gal/1000 ft³**.

Pressure drop across the venturi is then calculated to be:

$\Delta P = 5 \times 10^{-5}\ (272\ \text{ft/s})^2\ (14\ \text{gal}/1000\ \text{ft}^3) = 51.8$ inches of water

For 80% removal of 0.3 μm particles the calculations above are repeated using a C value = 1.5 @ 70 °F. For E = 80%, a value of **R = 16 gal/1000 ft³** is obtained, with a pressure drop of **ΔP = 59.2 inches of water.**

APCE-4 Solution.

a. The following plot of outlet concentration versus sample volume treated can be used to determine the breakthrough loading for required TCE removal.

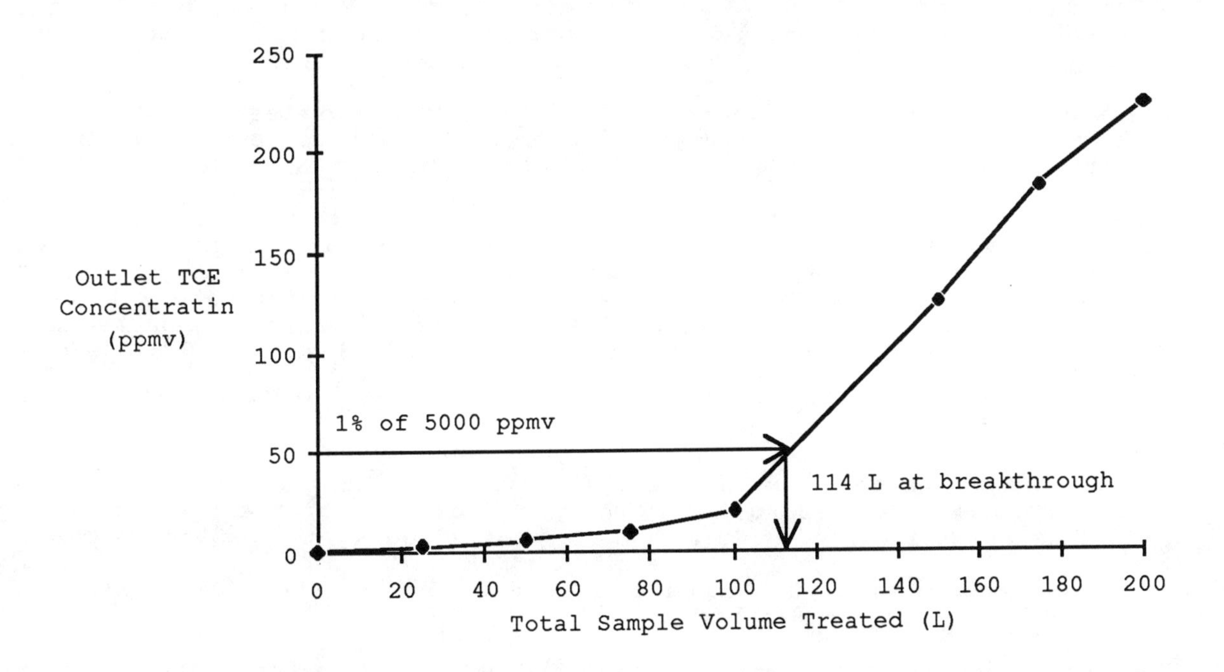

From this relationship, the loading of 114 L at 5000 ppmv TCE on 100 g of carbon is calculated as follows:

Molar volume of gas at 20 °C is: (293 K/273 K) (22.4 L) = **24.04 L/gmol**

5000 ppmv = (5 mmol TCE)/(24.04 L) = (658 mg)/(24.04 L) = **27.4 mg/L**

Therefore 114 L @ 27.4 mg/L gives a **mass loading of 3124 mg TCE**

The breakthrough loading from the experimental data was therefore 0.31 g TCE/g carbon for 97% TCE removal.

b. One drain-and-fill cycle displaces a volume of:

Volume = π d^2 h/4 = π $(3\ m)^2$ (4 m)/4 = **28.27 m^3**

At a concentration of 5000 ppmv, the displaced gas contains (@ 20 °C) 5 mmol TCE/24.04 L = 20.88 mg TCE/L or **20.88 g/m^3**

Therefore one volume displacement would involve:

$(21.88\ g/m^3)$ $(28.27\ m^3)$ = **588 g TCE**

The capacity of 6 kg of carbon is:

(6 kg) (0.31 kg TCE/kg carbon) = **1.86 kg TCE**

APCE-4 Solution (Continued).

At the rate of 0.619 kg per cycle, the carbon has a capacity of 1.86/0.508 = 3 **complete cycles without breakthrough.**

c. To go one full cycle beyond the breakthrough is the equivalent of loading another 588 g of TCE onto the carbon. Referring back to the breakthrough curve, 588 g TCE/6 kg carbon = 0.098 g TCE/g carbon = 980 mg TCE/g carbon.

Loading at 5000 ppmv = 27.4 mg/L, so 980 mg TCE requires 35.8 L beyond breakthrough. At 114 + 35.8 = 149.8 L, the outlet concentration is 125 ppmv, or 125/5000 = 0.025 = **2.5% breakthrough.**

For one extra cycle, an additional 28.27 m^3 would be displaced with the 5000 ppmv concentration. The corresponding discharge from the carbon would begin at 1% of 5000 ppmv and end at approximately 2.5% of 5000 ppmv. This portion of the breakthrough curve is approximately linear, so the 28.27 m^3 may be assumed to be discharged with an average concentration of 1.8% of 5000 ppmv = (0.018) (20.8 g/m^3) = **0.375 g/m^3**

Therefore, the total mass of TCE discharged = (0.375 g TCE/m^3) (28.27 m^3) = **10.6 g TCE from the additional cycle.**

d. One tank of waste represents 28.27 m^3 which is 28 wt% TCE. The waste has a density of 1.13 kg/L = 1130 kg/m^3. Therefore, the total mass of TCE to be incinerated = (0.28) (28.27 m^3) (1130 kg/m^3) = **8950 kg TCE.**

At 99.99% destruction, the maximum TCE emission from the incinerator is (0.0001) (8950 kg) = 895 g.

The excess discharge caused by running the activated carbon one cycle past breakthrough is 11.13 g TCE, which is 11.13/895 = 0.012 = 1.2% of the total permitted emission for the incineration of one entire tank of waste.

It should be noted that this fraction will rise rapidly as breakthrough is exceeded through a larger number of cycles.

APCE-5 Solution.

First, the frequency of cleaning is calculated from a rearrangement of the equation presented in the problem statement:

$t = (\Delta P - 0.2\ v)/(5\ W_i\ v^2)$

$= [8 \text{ inches of water} - 0.2\ (10 \text{ ft/min})]/[5\ (5 \text{ gr/ft}^3)/(7000 \text{ gr/lb})\ (10 \text{ ft/min})^2]$

$= 16.8 =$ **17 min**

Removal efficiencies and pressure drops are calculated next using the appropriate equations in the problem statement. These results are summarized in the table that follows:

APCE-5 Solution (Continued).

		Discharge Concentration (gr/ft^3)		Pressure Drop (inches Water)	
t (min)	E (%)	5 gr/ft^3	2 gr/ft^3	5 gr/ft^3	2 gr/ft^3
1	89	0.53	0.21	2.36	2.14
2	91	0.46	0.18	2.71	2.29
3	92	0.39	0.16	3.07	2.43
4	93	0.34	0.14	3.43	2.57
5	94	0.29	0.12	3.79	2.71
6	95	0.25	0.10	4.14	2.86
7	96	0.21	0.09	4.50	3.00
8	96	0.18	0.07	4.86	3.14
9	97	0.16	0.06	5.21	3.29
10	97	0.14	0.05	5.57	3.43
11	98	0.12	0.05	5.93	3.57
12	98	0.10	0.04	6.29	3.71
13	98	0.09	0.03	6.64	3.86
14	99	0.07	0.03	7.00	4.00
15	99	0.06	0.03	7.36	4.14
16	99	0.05	0.02	7.71	4.29
17	99	0.05	0.02	8.07	4.43
	Average	0.19	0.08		

APCE-6 Solution.

The humidity of the gases leaving the cooling tower at 100 °F is obtained from a psychrometric chart found in Appendix A, yielding a value of 0.044 lb H_2O/lb dry air.

The humidity of the gases entering the cooling tower is calculated as follows:

$$H = \frac{0.12\ (18\ \text{lb/lbmol}\ H_2O)}{0.1\ (44\ \text{lb/lbmol}\ CO_2) + 0.73\ (28\ \text{lb/lbmol}\ N_2) + 0.05\ (32\ \text{lb/lbmol}\ O_2)}$$

= **0.082 lb H_2O/lb dry air**

Therefore, the amount of water removed in the cooling process = 0.082 - 0.043 = **0.039 lb H_2O/lb dry air**

The gas entering the spray tower has the following volume and mass percent composition:

Compound	Volume %	Weight %
CO_2	10	15.4
H_2O	12	7.6
N_2	73	71.5
O_2	5	5.5

The dry gas mass flow rate = 5000 lb/h (1 - 0.076) = 4620 lb/h

The total mass of water removed from the flue gas

= 4620 lb/h (0.039 lb H2O/lb dry air) = **180.2 lb H_2O/h**

APCE-7 Solution.

On a mass basis, the reactant and product quantities from the neutralization equations are as follows:

$$HCl + 1/2\ Ca(OH)_2 \rightarrow 1/2\ CaCl_2 + H_2O$$
$$36.5\ lb + 0.5\ (74\ lb) \rightarrow 0.5\ (111\ lb) + 18\ lb$$

$$SO_2 + Ca(OH)_2 \rightarrow CaSO_3 + H_2O$$
$$64\ lb + 74\ lb \rightarrow 120\ lb + 18\ lb$$

$$CO_2 + Ca(OH)_2 \rightarrow CaCO_3 + H_2O$$
$$44\ lb + 74\ lb \rightarrow 100\ lb + 18\ lb$$

The gas flow rate is changed to standard conditions, i.e., 32 °F and 1 atm, as follows:

(16,000 acfm) [(32 + 460 °R)/(600 + 460 °R)] = **7400 scfm**

The lime requirement for this flow rate is calculated as follows:

[(7400 scfm)/(359 ft^3/lbmol)] {[(1500 ppm HCl)/(10^6 parts/million parts)] (0.999 removal) (0.5 mol lime/mol HCl) + [(470 ppm SO_2)/(10^6 parts/million parts)] (0.75 removal) (1.0 mol lime/mol SO_2) + (0.13 mole fraction CO_2) (0.01 removal) (1.0 mol lime/mol CO_2)}
= 0.05 lbmol lime/min

At two times the stoichiometric dose, the total mass per unit time required is: 0.05 lbmol lime/min (2) (60 min/h) (74 lb lime/lbmol)
= **440 lb lime/h**

The ash generated from the system is calculated as follows:

From the acid gases:

[(7400 scfm)/(359 ft^3/lbmol)] {[(1500 ppm HCl)/(10^6 parts/million parts)] (0.999 removal) (111 lb $CaCl_2$/lbmol) (0.5 lbmol $CaCl_2$/mol HCl) + [(470 ppm SO_2)/(10^6 parts/million parts)] (0.75 removal) (120 lb $CaSO_3$/lbmol) (1.0 mol $CaSO_3$/mol SO_2) + (0.13 mole fraction CO_2) (0.01 removal) (100 lb $CaCO_3$/lbmol) (1.0 mol $CaCO_3$/mol CO_2)} = **5.3 lb ash/min**

Excess lime = 0.05 lbmol lime/min (74 lb lime/lbmol) = **3.7 lb/min**

From particulate removal:

[(7400 scfm) (2.2 - 0.01 gr/scf)]/(7000 gr/lb) = **2.32 lb/min**

Total ash produced then = 5.3 + 3.7 + 2.32 lb/min = **11.2 lb/min = 674 lb/h**

APCE-8 Solution.

The superficial gas velocity is converted to SI units:
(2.0 ft/min)/(3.3 ft/m) = **0.61 m/min**

Particulate concentration to the fabric filter:
(5 gr/acf) (0.0648 g/gr) (1 ft^3/0.02832 m^3) = **11.44 g/m³**

The particulate mass loading versus filter drag are plotted as shown below to yield the coefficients K_1 and K_2 as shown below.

APCE-8 Solution (Continued).

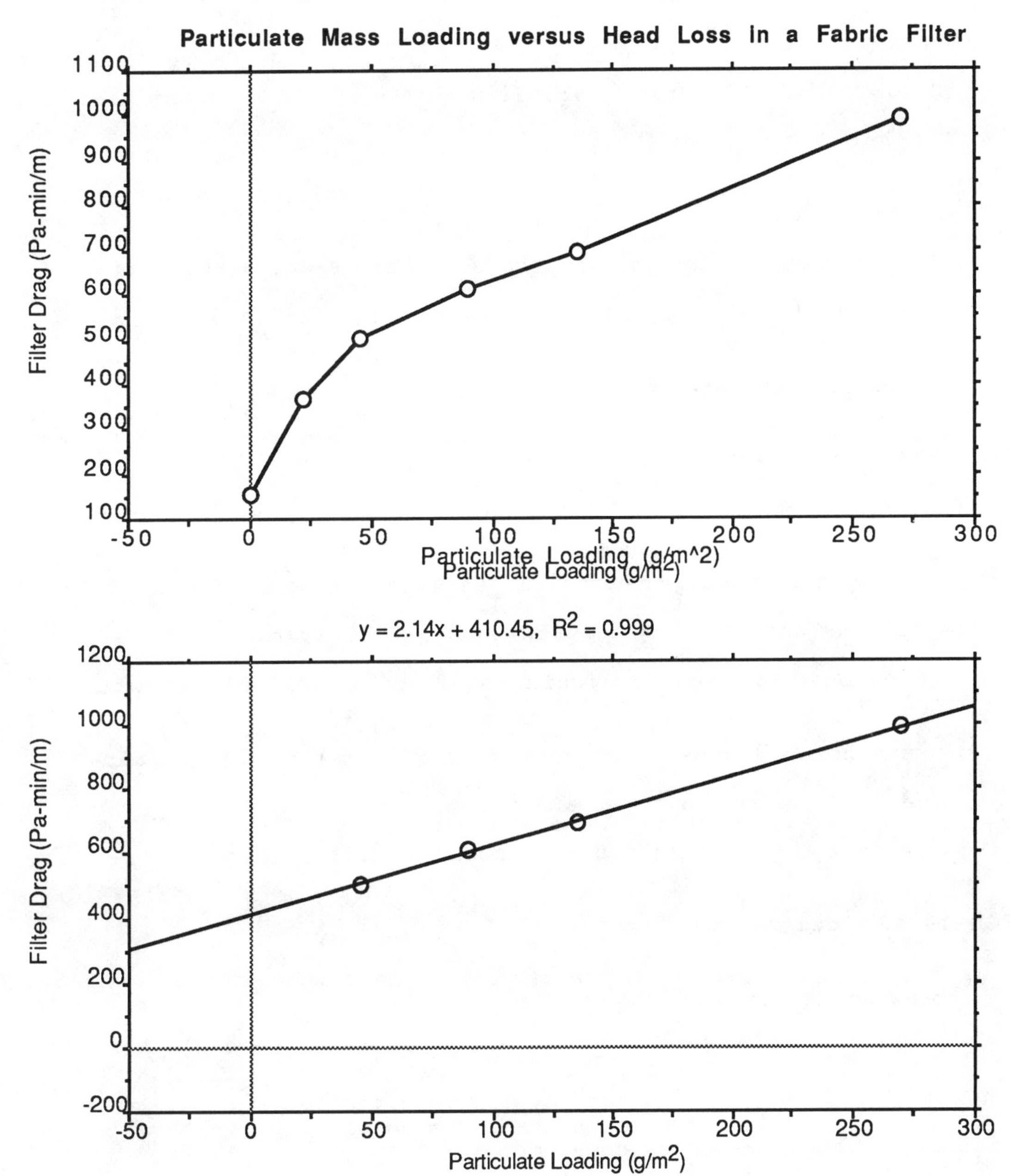

From the last figure, K_1 = 410 Pa-min/m, and K_2 = 2.14 Pa-min-m/g.

Rearranging the expression given in the problem yields an expression for filtering t as shown below:

$$t = S/(K_1 + K_2 C_p V_g) = \Delta P/[V_g (K_1 + K_2 C_p V_g)]$$

APCE-8 Solution (Continued).

Solving for t yields:

$$t = (2000 \text{ Pa})/\{(0.61 \text{ m/min}) [410 \text{ Pa-min/m} + (2.14 \text{ Pa-min-m/g}) (11.44 \text{ g/m}^3) (0.61 \text{ m/min})]\}$$
$$= \mathbf{7.7 \text{ min}}$$

APCE-9 Solution.

a. The volumetric flow rate in the scrubber is calculated as:

$$Q = (40{,}000 \text{ lb/h}) (359 \text{ ft3/lb}) (1 \text{ h/60 min}) [(460 + 275 \text{ °R})/(460 + 32 \text{ °R})]$$
$$= \mathbf{12{,}769 \text{ acfm}}$$

Using the approximate approach:

$$A = 0.2 [(40{,}000 \text{ lb/h}) (1 \text{ h/3600 s})]/\{(28 \text{ lb/359 ft}^3) [(460 + 32 \text{ °R})/(460 + 2200 \text{ °R})]\} = 0.2 (11.11 \text{ lb/s})/(0.0144 \text{ lb/ft}^3) = \mathbf{154 \text{ ft}^2}$$

Therefore, $D = [4 (154 \text{ ft}^2)/\pi]^{0.5}$ = **14 ft**, and the tower height Z = 2 D = **28 ft.**

b. The scrubber tower superficial gas velocity is calculated as follows:

$$V = (12{,}769 \text{ acfm}/154 \text{ ft}^2) (1 \text{ min}/60 \text{ s}) = \mathbf{2.8 \text{ ft/s}}.$$

c. The amount of recycle liquid required at a liquid to gas ratio of 25 gal/1000 acf is found to be:

Volume water = (25 gal/1000 acf) (12,769 acfm) = **319 gal/min.**

d. The amount of HCl removed is:

HCl removal = (200 lb/h) (0.99 removal) = **198 lb/h.**

APCE-10 Solution.

a. For dilute solutions, Henry's law describes the phase equilibrium between HCl in a liquid and gas phase as:

$$x_i = y_i/m = 0.02/0.2 = 0.1$$

where x_i = the equilibrium liquid concentration
y_i = mole fraction in the incoming gas
m = Henry's law constant.

Since 99.5% of the HCl is removed, the outlet gas stream concentration y_2 is:

$$y_2 = y_1 (1 - n)/[1 - y_1 + (1 - n) (y_2)]$$

where n = fractional scrubber efficiency
y_1 = the inlet gas concentration

Substituting and solving yields:

$$y_2 = (0.005) (0.02)/[1 - 0.02 + (0.005) (0.02)] = \mathbf{0.0001}$$

APCE-10 Solution (Continued).

Defining x_2 as the inlet liquid liquid concentration, the minimum liquid to gas ratio that can be used is given as:

$(L/G)_{min} = (y_1 - y_2)/(x_1 - x_2) = (0.02 - 0.0001)/(0.1 - 0) =$ **0.199**

Since the specified liquid flow is 150% more than the minimum, the actual liquid-to-gas ratio is:

$(L/G)_{actual} = 0.199\ (2.5) =$ **0.498**

b. At 50% recycle, the inlet concentration is:

$x_2 = x_1\ (0.5) + 0\ (0.5) = 0.1\ (0.5) =$ **0.05**

The minimum recycle is then:

$(L/G)_{min} = (0.02 - 0.0001)/(0.1 - 0.05) =$ **0.398**

Since the specified liquid flow is 150% more than the minimum, the actual liquid to gas ratio is:

$(L/G)_{actual} = 0.398\ (2.5) =$ **0.995**

APCE-11 Solution.

A 1 h basis is chosen for carrying out necessary calculations. The outlet concentration for the original design was:

$y_2 = (0.01\ y_1)/[(1 - y_1) + (0.01\ y_1)] = (0.01)(0.02)/[(1 - 0.02) + 0.01\ (0.02)] =$ **0.0002**

The number of transfer units can be calculated from this concentration as:

$N_{OG} = \ln\{[(y_1 - m\ x_2)/(y_2 - m\ x_2)](1 - 1/A) + 1/A\}/(1 - 1/A)$

Since liquid is pure, $x_2 = 0$, then

$(y_1 - m\ x_2)/(y_2 - m\ x_2) = 0.05/0.0002 = 250$

The absorption factor, A, is calculated as,

$A = L/(m\ G) = \{(3690\ lb/h)/[(0.2)\ (5000\ lb/h)]\}\ (29\ lb/lbmol\ air)/(18\ lb/lbmol\ water) = 5.94$

The number of transfer units is then:

$N_{OG} = \ln[250\ (1 - 1/5.94) + 1/5.94)]/(1 - 1/5.94) =$ **5.5**

The height of the packed column is given by the product of the number of transfer units and the height of each unit as:

$Z = N_{OG}\ H_{OG} = 5.5\ (2.5\ ft) =$ **13.8 ft**

Therefore the existing column is adequate.

The liquid-to-gas ratio under the new operating conditions based on the minimum ratio is calculated by first calculating the $(L/G)_{min}$.

APCE-11 Solution (Continued).

The equilibrium liquid concentration under the new operating conditions is first calculated as follows:

$x_1^* = y_1/m = 0.05/0.2 = 0.25$

Then,

$(L/G)_{min} = (y_1 - y_2)/(x_1^* - x_2) = (0.05 - 0.002)/(0.25 - 0) = \mathbf{0.20}$

With and $(L/G)_{actual} = 1.188$, it represents 494% of the minimum design value, essentially equivalent to the original design of 500% of $(L/G)_{min}$.

APCE-12 Solution.

The volumetric flow rate of the inlet gas = $(4000\ lb/h)/(0.075\ lb/ft^3)$ = **53,333 ft^3/h**

The HCl flow rate = $(53{,}333\ ft^3/h)\ (0.05)$ = **2667 ft^3/h**

The inert gas flow rate = 53,333 - 2667 = **50,666 ft^3/h = 3800 lb/h**

Therefore, the column exhaust gas contains 4 lb/h HCl and 3800 lb/h inert gases.

The density of HCl @ 70 °F = 0.094 lb/ft^3

The mole fraction of HCl in the exhaust gas is:
$= [4\ lb/(0.094\ lb/ft^3)]/[(3800\ lb)/(0.075\ lb/ft^3)] = \mathbf{0.00084}$

and the equilibrium mole fraction in the liquid is:
$= x_1^* = 0.05/0.2 = \mathbf{0.25}$

The minimum liquid-to-gas flow rate is calculated as:

$(L/G)_{min} = (y_1 - y_2)/(x_1^* - x_2) = (0.05 - 0.00084)/(0.25 - 0) = \mathbf{0.197}$

A (L/G) of two times the minimum = 0.198 (2) = **0.396**

$N_{OG} = \ln\{[(y_1 - m\ x_2)/(y_2 - m\ x_2)](1 - 1/A) + 1/A\}/(1 - 1/A)$

$(y_1 - m\ x_2)/(y_2 - m\ x_2) = 0.05/0.00084 = \mathbf{59.52}$

$A = L/(m\ G) = 0.396/0.2 = \mathbf{1.98}$

Substituting and solving for the number of transfer units yields N_{OG} = **4.07**

The height of the packed column is calculated from the number of transfer units and the height of each unit as:

$Z = N_{OG}\ H_{OG} = 4.07\ (2.5\ ft) = \mathbf{10.2\ ft}$

The tower diameter is calculated from the gas velocity at 60% of flooding velocity, i.e., $V = 0.6\ (500\ lb/h\text{-}ft^2)$ = **300 $lb/h\text{-}ft^2$**

The cross-sectional area of the tower = $(4000\ lb/h)/(300\ lb/h\text{-}ft^2)$ = **13.3 ft^2**

And the tower diameter = $[(13.3\ ft)\ (4)/\pi]^{0.5}$ = **4.12 ft**

APCE-13 Solution.

a. The neutralization reactions which occur in the packed bed absorber and reactant and product quantities on a mass basis are as follows:

$$HCl + NaOH \rightarrow NaCl + H_2O$$
$$36.5 \text{ lb} + 40 \text{ lb} \rightarrow 58.5 \text{ lb} + 18 \text{ lb}$$

$$SO_2 + 2\ NaOH \rightarrow Na_2SO_3 + H_2O$$
$$64 \text{ lb} + 2\ (40 \text{ lb}) \rightarrow 126 \text{ lb} + 18 \text{ lb}$$

$$CO_2 + NaOH \rightarrow NaHCO_3$$
$$44 \text{ lb} + 40 \text{ lb} \rightarrow 84 \text{ lb}$$

The gas flow rate is changed to standard conditions, i.e., 32 °F and 1 atm, as follows:

(10,00 acfm) [(32 + 460 °R)/(180 + 460 °R)] = **7700 scfm**

The NaOH requirement for this flow rate is calculated as follows:

[(7700 scfm)/(359 ft^3/lbmol)] {[(1500 ppm HCl)/(10^6 parts/million parts)] (0.999 removal) (1 mol NaOH/mol HCl) + [(470 ppm SO_2)/(10^6 parts/million parts)] (0.65 removal) (2.0 mol NaOH/mol SO_2) + (0.13 mole fraction CO_2) (0.01 removal) (1.0 mol NaOH/mol CO_2)} = **0.073 lbmol NaOH/min**

On a mass per unit time basis, the requirement is: 0.073 lbmol NaOH/min (60 min/h) (40 lb NaOH/lbmol) = **175.2 lb NaOH/h**

On a volume per unit time basis, the requirement is: 175.2 lb NaOH/h/[(1.31) (8.34 lb water/gal)] = 16 gal/h. Since the NaOH solution is only 23% by weight NaOH, the actual volumetric requirement = 16/0.23 = **69.72 gal/h**

b. The purge flow rate is calculated based on the solids generated from the neutralization reactions presented above. The dissolved solids generated as calculated as:

[(7700 scfm)/(359 ft^3/lbmol)] {[(1500 ppm HCl)/(10^6 parts/million parts)] (0.999 removal) (58.5 lb NaCl/lbmol) (1 lbmol NaCl/mol HCl) + [(470 ppm SO_2)/(10^6 parts/million parts)] (0.65 removal) (126 lb $NaSO_3$/lbmol) (1.0 mol $CaSO_3$/mol SO_2) + (0.13 mole fraction CO_2) (0.01 removal) (84 lb $NaHCO_3$/lbmol) (1.0 mol $NaHCO_3$/mol CO_2)} = **5.03 lb TDS/min**

The volumetric flow rate = (5.03 lb TDS/min)/[(8.34 lb water/gal) (0.05)] = **12.06 gal/min = 724 gal/h**

APCE-14 Solution.

a. The neutralization reactions which occur in the spray dryer and reactant and product quantities on a mass basis are as follows:

$$2\ HCl + 1.1\ Ca(OH)_2 \rightarrow CaCl_2 + 2\ H_2O + 0.1\ Ca(OH)_2$$
$$73 \text{ lb} + 1.1\ (74 \text{ lb}) \rightarrow 111 \text{ lb} + 2\ (18 \text{ lb}) + 0.1\ (74 \text{ lb})$$

$$SO_2 + 0.5\ O_2 + 1.3\ Ca(OH)_2 \rightarrow CaSO_4 + H_2O + 0.3\ Ca(OH)_2$$
$$64 \text{ lb} + 0.5\ (32 \text{ lb}) + 1.3\ (74 \text{ lb}) \rightarrow 120 \text{ lb} + 18 \text{ lb} + 0.3\ (74 \text{ lb})$$

APCE-14 Solution (Continued).

The gas flow rate is changed to standard conditions, i.e., 32 °F and 1 atm, as follows:

(10,00 acfm) [(32 + 460 °R)/(750 + 460 °R)] = **4066 scfm**

The mass flow rate of HCl produced in the incinerator is calculated as follows:

[(4066 scfm)/(359 ft^3/lbmol)] (0.02 mol fraction HCl) (36.45 lb HCl/lbmol) = **8.26 lb/min = 495 lb HCl/h**

To determine which HCl control requirement is less strict, the 99% removal criteria is checked.

Therefore 99% removal = 495 (0.01) = **4.95 lb/h.** This is the requirement that must be met.

The mass flow rate of SO_2 produced in the incinerator is calculated as follows:

[(4066 scfm)/(359 ft^3/lbmol)] [(350 ppm SO_2)/(10^6 parts/million parts)] (64 lb SO_2/lbmol) = **0.25 lb/min = 15.2 lb SO_2/h**

These mass production rates correspond to the following mol production rates:

HCl: (495 lb HCl/h)/(36.45 lb HCl/lbmol) = **13.6 lbmol HCl/h**
SO_2: (15.2 lb SO_2/h)/(64 lb SO_2/lbmol) = **0.28 lbmol SO_2/h**

The $Ca(OH)_2$ requirement for the system is calculated based on the stoichiometric equations presented above:

1 lbmol SO_2 requires 1.3 lbmol $Ca(OH)_2$; so for SO_2, the total requirement = 0.28 lbmol SO_2/h (1.3 lbmol $Ca(OH)_2$) = **0.364 lbmol $Ca(OH)_2$/h**

1 lbmol HCl requires 0.55 lbmol $Ca(OH)_2$; so for HCl, the total requirement = 13.6 lbmol SO_2/h (0.55 lbmol $Ca(OH)_2$) = **7.48 lbmol $Ca(OH)_2$/h**

On a mass per unit time basis, the total requirement is: (0.364 + 7.48 lbmol $Ca(OH)_2$/h) (74 lb $Ca(OH)_2$/lbmol) = **580.5 lb $Ca(OH)_2$/h**

b. The spent solids generated include $CaCl_2$, $Ca(OH)_2$, and $CaSO_4$, and are calculated from the stoichiometric expressions presented above:

[(4066 scfm)/(359 ft^3/lbmol)] {(0.02 mol fraction HCl) (0.99 removal) (111 lb $CaCl_2$/lbmol) (0.5 lbmol $CaCl_2$/lbmol HCl) + [(350 ppm SO_2)/(10^6 parts/million parts)] (0.70 removal) (120 lb $CaSO_4$/lbmol) (1.0 lbmol $CaSO_4$/lbmol SO_2) + (0.02 mol fraction HCl) (0.99 removal) (74 lb $Ca(OH)_2$/lbmol) (0.05 lbmol $Ca(OH)_2$/lbmol HCl) + [(350 ppm SO_2)/(10^6 parts/million parts)] (0.70 removal) (74 lb $Ca(OH)_2$/lbmol) (0.3 lbmol $Ca(OH)_2$/lbmol SO_2)} = **13.9 lb dry solids/min = 835 lb dry solids/h**

Since the solids are 10% by weight water, the total sludge mass production rate = (835 lb/h) (1.1) = **918 lb sludge/h**

APCE-15 Solution.

a. The amount of mercury bound to the fly ash which is captured in the ESP is calculated based on the mass of ash leaving the stack. This mass is calculated as follows:

Mass of ash out stack = (0.08 gr/dscf) (1 lb/7000 gr) (20,000 cfm) (60 min/h) (24 h/d) = **329 lb ash/d**

The mass exiting the stack represents 0.5% of the ash collected in the ESP, i.e.,

Mass in the ESP = (329 lb/d)/0.0005 = **65,829 lb ash/d**

The amount of mercury bound to the fly ash in the ESP is calculated based on the mercury content of the fly ash:

Mercury in the ESP = (65,829 lb ash/d) (2.42 µg mercury/g ash) (1 g/1,000,000 µg) (454 g/lb) = **72.3 lb/d**

b. The amount of mercury leaving the stack as a vapor and with the fly ash is calculated as:

Mercury leaving as a vapor = (20,000 dscfm) (0.02831 m3/ft3) (0.3 mg mercury/dscm) (1 g/1,000 mg) (60 min/h) (24 h/d)/(454 g/lb) = **0.54 lb/d**

Mercury leaving with fly ash = (329 lb ash/d) (2.42 µg mercury/g ash) (1 g/1,000,000 µg) (454 g/lb) = **0.36 lb/d**

Total mercury emission = 0.54 + 0.36 = **0.90 lb/d**

APCE-16 Solution.

a. The ESP's minimum required particulate mass collection efficiency is based on the required particulate discharge requirement of 0.08 gr/dscf corrected to 50% EA. Using the particulate loading rate and this performance requirement, the minimum collection efficiency is:

E_{min} = (5.0 - 0.08 gr/dscf)/(5.0 gr/dscf) = 0.984 = 98.4%

b. The minimum collection area of the ESP to achieve the required particulate mass collection efficiency can be calculated using the analytic expression for ESP particle collection efficiency given in the problem statement.

First the particle migration velocity is calculated as follows:

$u = (C\ d_p\ \varepsilon_o\ K\ E^2)/(3\ \mu_g)$ = [1.0 (d_p) (8.85 x 10^{-2} c/v-m) (5) (300,000 v/m)^2]/[3 (2.48 x 10^{-5} kg/m-s)] = **(d_p) (5.35 x 10^4)**

For d_p = 0.3 µm, u = (0.3 x 10^{-6}) (5.35 x 10^4) = 1.61 x 10^{-2} m/s = **0.96 m/min**

For d_p = 5.0 µm, u = (5.0 x 10^{-6}) (5.35 x 10^4) = 2.68 x 10^{-1} m/s = **16.06 m/min**

These results are then used to calculate a required collection area to yield 98.4% particle removal using the ESP particle removal efficiency equation.

APCE-16 Solution (Continued).

Substituting into the efficiency equation yields:

$$0.984 = \{0.3\,[1 - \exp(-A\,(0.96\ \text{m/min})/(5000\ \text{m}^3/\text{min})] + 0.7\,[1 - \exp(-A\,(16.06\ \text{m/min})/(5000\ \text{m}^3/\text{min})]\}$$

A can be solved by making an initial assumption that the efficiency of 5.0 μm particles = 100% (second term of the above equation drops out), and then solving for an initial value of A. The final value of A is then determined by iteration using the updated values of A until the values of A converge. If the process is continued to convergence, the required collector area is found to be: **A = 15,267 m^2.**

CHAPTER EIGHT

Risk Analysis

RA-1 Solution.

a. High CO can indicate incomplete combustion. Emissions of both POHCs and PICs are likely to be higher under such conditions. Immediate corrective action is necessary.

b. Low temperature in secondary combustion chamber means that DREs for POHCs may not be achieved because of inadequate combustion temperature. Corrective action should be immediate.

c. High combustion gas flow rate means that residence times within the combustion zone in the incinerator will be lower, reducing the net POHC conversion and potentially producing an effluent not meeting the required DRE.

d. Low scrubber water flow rate means that HCl removal will likely be below design specifications. It may also mean that 99% HCl (or $\leq$ 4 lb/h) required HCl emission rate may not be continuously met. There is also a danger to downstream equipment if corrosive gases pass through the scrubber in excessive concentrations. Corrective action is usually immediate.

e. Low quench water flow rate will mean inadequate quenching which can endanger downstream equipment (particularly plastics) because of excessive flue gas temperatures. Immediate corrective action is necessary.

f. High pressure in either combustion chamber should not occur since the entire system is operated under negative pressure from the ID fan. High internal incinerator pressures can develop if high Btu wastes are rapidly loaded to the incinerator. If pressure builds in a rotary kiln, emissions can occur from the seals on the kiln, and these emissions will not be thoroughly combusted nor vented from the incinerator area. Unvented seal leakage and high pressure at sufficient levels can lead to exposure and explosion hazards, with resulting danger to workers and damage to equipment.

g. High temperature in either or both combustion chambers can endanger the refractory integrity and thereby the incinerator shell, thus resulting in worker safety (explosion) concerns. High incinerator temperatures can also affect downstream equipment as indicated for low quench water flow.

h. Flame loss in either combustion chamber will allow combustible gases to accumulate within the incinerator to possible explosive levels, resulting in increased risks to both personnel and equipment. Emissions would be unacceptably high during flame-out conditions and immediate corrective action is necessary.

RA-2 Solution.

Assume that the partial pressure of the gas in the tank is given by its vapor pressure:

$$\text{Partial Pressure} = \frac{95}{760} = 0.125 \text{ atm @ } 75\ ^{\circ}\text{F}$$

The concentration of benzene in the vent gas at 75°F based on the ideal gas law is calculated as:

RA-2 Solution (Continued).

$$\text{Concentration} = \frac{n\ M}{V} = \frac{P\ M}{R\ T} = \frac{0.125\ \text{atm}\ (78\ \text{lb/lbmol})}{0.7302\ \text{atm-ft}^3/\text{lbmol-}^\circ\text{R}\ (460+75^\circ\text{R})}$$

$$= 0.025\ \text{lb/ft}^3\ @\ 75^\circ\text{F}$$

The emission rate from the vent is:

$$\text{Emission Rate} = 10{,}000\ \text{gal/d}\ (1\ \text{ft}^3/\text{gal})\ (0.025\ \text{lb/ft}^3) = \mathbf{33.5\ lb/d}$$

Emission rate from the incinerator:

$$\text{Emission Rate} = 10{,}000\ \text{gal/d}\ (\text{S.G.})\ (8.34\ \text{lb/gal})\ (\text{Efficiency}) =$$
$$= 10{,}000\ \text{gal/d}\ (0.87)\ (8.34)\ (1-0.9999) = \mathbf{7.3\ lb/d}$$

The vent from the tank discharges more hydrocarbons than the incinerator. Controls are required on the tank for these "fugitive" emissions, and usually take the form of a nitrogen blanket over the tank contents along with a carbon adsorber on the vent line to reduce vent emissions.

RA-3 Solution.

a. Any number of scenarios might be proposed, such as:

1. A spill of waste from the tank or feed system.
2. A blow-back of waste upon over-charging to the kiln.
3. Excess air feed control problems.
4. Reduction or failure of flow in the fuel line.
5. Reduction or failure of flow in the quench/absorber water line.
6. Power loss.
7. Faulty emergency vent stack release.
   etc.

Note that all of the above scenarios are related to incinerator operations and POHC emissions.

b. Although other methods are available, the best approach for developing a probability estimate would be a fault tree analysis.

c. Given a POHC emission scenario and the probability of occurrence, the main concern would be the estimation of a maximum possible acute exposure, and determining where the exposure would take place. Dispersion modeling would be the logical choice in exposure assessment and prediction. For acute exposure, a carcinogenic risk assessment would likely be inappropriate. Rather a comparison of maximum values with short-term exposure limits would make the most sense. Event trees are often employed here.

d. For any risk scenario, the overall risk can be reduced by:

1. Taking steps to eliminate the scenario by modifying the process.
2. Reducing the probability of the scenario by identifying key components from the fault tree analysis and modifying them to lower overall risk probability levels.
3. Taking steps to reduce the consequences, e.g., establishing a larger buffer zone, reducing acute exposure, reducing population at risk, etc.

RA-4 Solution.

a. Assume fine particulates discharged from the stack behave as gases so that Gaussian Dispersion modeling can be used to estimate downwind concentrations. Based on the Holland equation, the effective stack height from plume rise calculations is determined as:

$$\Delta H = \frac{(13 \text{ m/s})(0.75 \text{ m})}{2 \text{ m/s}} \left(1.5 + 0.0056\ (101 \text{ mPa})\ (0.75 \text{ m}) \left[\frac{315\text{-}288 \text{ K}}{315 \text{ K}}\right]\right)$$

$$= 7.49 \text{ m} = \mathbf{7.5\ m}$$

$$H = 20 \text{ m} + 7.5 \text{ m} = \mathbf{27.5\ m}$$

Dispersion calculations are made based on the Gaussian Dispersion equation given above for centerline, ground level concentrations where y = 0 and z = 0.

$Q = 20{,}000 \text{ m}^3\text{/h}\ (20 \text{ mg/m}^3)\ (1 \text{ g/103 mg}) = 400 \text{ g/h} = \mathbf{111.1\ mg/s}$

$$C(x,0,0;H) = \frac{Q}{\pi\ \sigma_y\ \sigma_z\ u} - \exp\left[-\frac{1}{2}\left(\frac{H}{\sigma_z}\right)^{0.5}\right]$$

For stability category F, this equation can be solved using the following X distances and σy and σz values:

x (km)	σy (m)	σz (m)	C (mg/m^3)
0.5	20	7.5	0.00062
0.8	30	12	0.0036
1.0	34	14	0.0054
1.5	50	18	0.0061
2.0	64	22	0.0057
3.0	100	28	0.0039
5.0	150	36	0.0024

Graphically, these data look as shown below:

The maximum concentration occurs at 1.5 km where the estimated particulate concentration is 0.0061 mg/m^3.

b. The suspended particulate matter from the incinerator is 0.6% of the tavern concentration, 2.2% of the church bingo area, and 7.1% of the restaurant environment.

Data have shown that cigarette smoke particulates cause lung cancer, yet one can only speculate about the human cancer risk from hazardous waste incinerator emissions because there is no evidence on their hazard to humans. However, animal evidence available would not support a premise that hazardous waste incinerator particulate discharges are several orders of magnitude more hazardous than cigarette smoke particulates on a mass basis. Consequently, point of plume impingement under conditions of poor meterological mixing appears to be the "safest" location compared to the other environments given above.

RA-4 Solution (Continued).

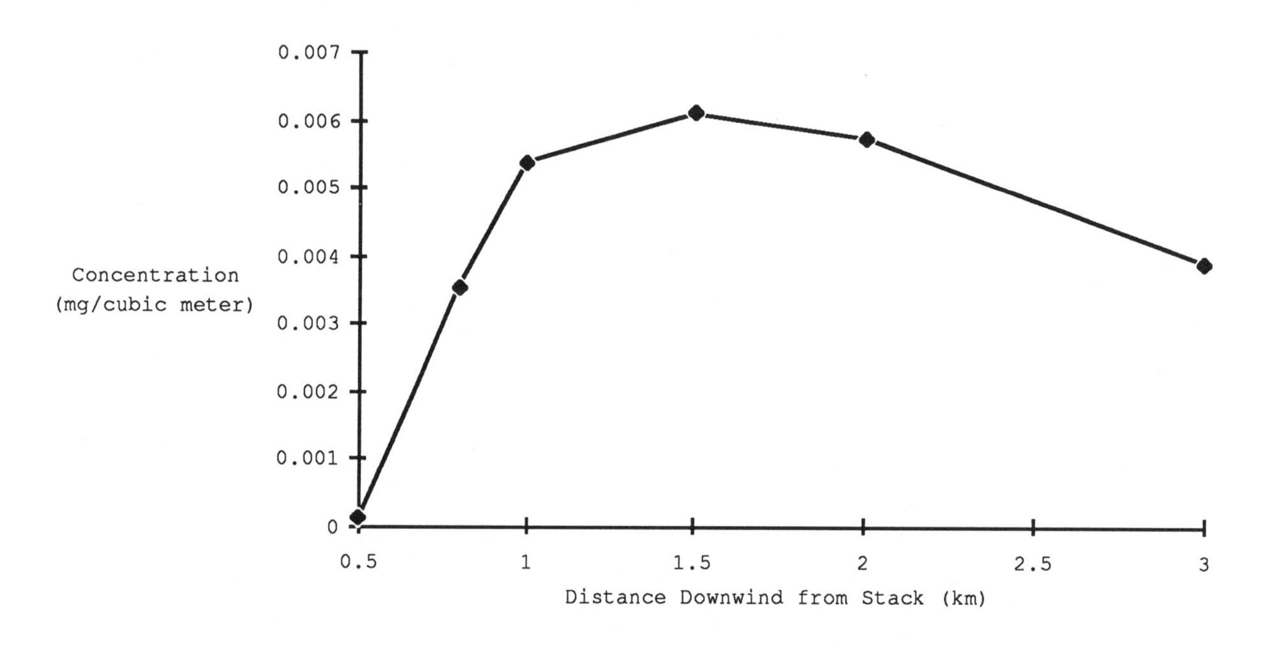

RA-5 Solution.

a. The maximum concentration will occur at ground level along the center of the gas plume at some point downwind at $x = x_{max}$ depending on meterological conditions. The "puff" dispersion model has the following formulation:

$$C(x,0,0;H) = \frac{2\,Q_r}{(2\pi)^{3/2}\,\sigma_x\,\sigma_y\,\sigma_z}\exp\left[-\frac{1}{2}\left(\frac{x-u\,t}{\sigma_x}\right)^2\right]\exp\left[-\frac{1}{2}\left(\frac{H}{\sigma_z}\right)^2\right]\exp\left[-\frac{1}{2}\left(\frac{y}{\sigma_y}\right)^2\right]$$

Centerline concentrations represent maximum levels, i.e., where y = 0, and where the time to receptor location x = u t. The complete equation above can be simplified for this maximum ground level, centerline concentrations to the following form:

$$C_{max}(x,0,0;H) = \frac{2\,Q_r}{(2\,\pi)^{3/2}\,\sigma_x\,\sigma_y\,\sigma_z}\exp\left[-\frac{1}{2}\left(\frac{H}{\sigma_z}\right)^2\right]$$

The derivative of C(x,0,0;H) with respect to x shows that the maximum concentration occurs were $\sigma_z = H/2^{-1/2}$. Ignoring plume rise, H = 10 m. Therefore $\sigma_z = 10/2^{-1/2}$ = **7.07 m ⇒ x = 160 m from Appendix J.**

At x = 160 m, **σy = σx = 12.5 m.** With an instantaneous release of 10 s of feed, and a waste feed rate of 7200 lb/h = 20 lb/s, the total release rate Q_r = 2 lb/s (10 s) = 20 lb.

RA-5 Solution (Continued).

$$C_{max}(160,0,0;10) = \frac{2\,Q_r\,2^{1/2}}{(2\,\pi)^{3/2}\,\sigma_x\,\sigma_y\,H}\,\exp\left[-\frac{1}{2}\left(\frac{H\sqrt{2}}{H}\right)^2\right]$$

$$= \frac{Q_r}{\pi\quad\sigma_y\,\sigma_x\,H}\,\exp^{-1}$$

Substituting appropriate values into the last equation presented above yields the maximum centerline concentration occurring at 160 m downstream from the puff release as:

$$C(160,0,0;10) = \frac{20\text{ lb}}{\pi^{1.5}\,(12.5\text{ m})\,(12.5\text{ m})\,(10\text{ m})\,(2.718)} = 8.457 \times 10^{-4}\text{ lb/m}^3$$

$$= 8.457 \times 10^{-4}\text{ lb/m}^3\;(453.6 \times 10^6\;\mu\text{g/lb}) = \mathbf{383{,}529\;\mu g/m^3}$$

At 50 m downwind distance, $\sigma y = \sigma x = 4.2$ m, $\sigma z = 2.5$ m

$$C(50,0,0;10) = \frac{2\;(20\text{ lb})}{(2\,\pi)^{1.5}\,(4.2\text{ m})\,(4.2\text{ m})\,(2.5\text{ m})}\;e^{-\frac{1}{2}\left(\frac{10\text{ m}}{2.5\text{ m}}\right)^2}$$

$$= 1.93 \times 10^{-5}\text{ lb/m}^3\;(453.6 \times 10^6\;\mu\text{g/lb}) = \mathbf{8763\;\mu g/m^3}$$

At 100 m downwind distance, $\sigma y = \sigma x = 8.0$ m, $\sigma z = 4.6$ m

$$C(100,0,0;10) = \frac{2\;(20\text{ lb})}{(2\,\pi)^{1.5}\,(8.0\text{ m})\,(8.0\text{ m})\,(4.6\text{ m})}\;e^{-\frac{1}{2}\left(\frac{10\text{ m}}{4.6\text{ m}}\right)^2}$$

$$= 8.12 \times 10^{-4}\text{ lb/m}^3\;(453.6 \times 10^6\;\mu\text{g/lb}) = \mathbf{368{,}392\;\mu g/m^3}$$

At 1000 m downwind distance, $\sigma y = \sigma x = 68.0$ m, $\sigma z = 32$ m

$$C(1000,0,0;10) = \frac{2\;(20\text{ lb})}{(2\,\pi)^{1.5}\,(68\text{ m})\,(68\text{ m})\,(32\text{ m})}\;e^{-\frac{1}{2}\left(\frac{10\text{ m}}{32\text{ m}}\right)^2}$$

$$= 1.63 \times 10^{-5}\text{ lb/m}^3\;(453.6 \times 10^6\;\mu\text{g/lb}) = \mathbf{7415\;\mu g/m^3}$$

At 4000 m downwind distance, $\sigma y = \sigma x = 245$ m, $\sigma z = 78$ m

RA-5 Solution (Continued).

$$C(4000,0,0;10) = \frac{2\ (20\ \text{lb})}{(2\ \pi)^{1.5}(245\ \text{m})(245\ \text{m})(78\ \text{m})}\ e^{-\frac{1}{2}\left(\frac{10\ \text{m}}{78\ \text{m}}\right)^2}$$

$$= 5.38 \times 10^{-7}\ \text{lb/m}^3\ (453.6 \times 10^{6}\ \mu\text{g/lb}) = \mathbf{244\ \mu g/m^3}$$
$$= 2.39 \times 10^{3}\ \mu\text{g/m}^3$$

Again, the location of maximum concentration is at x = 160 m. The relationship of the variation of centerline concentration downwind from the puff release versus downwind distance is presented in the following figure.

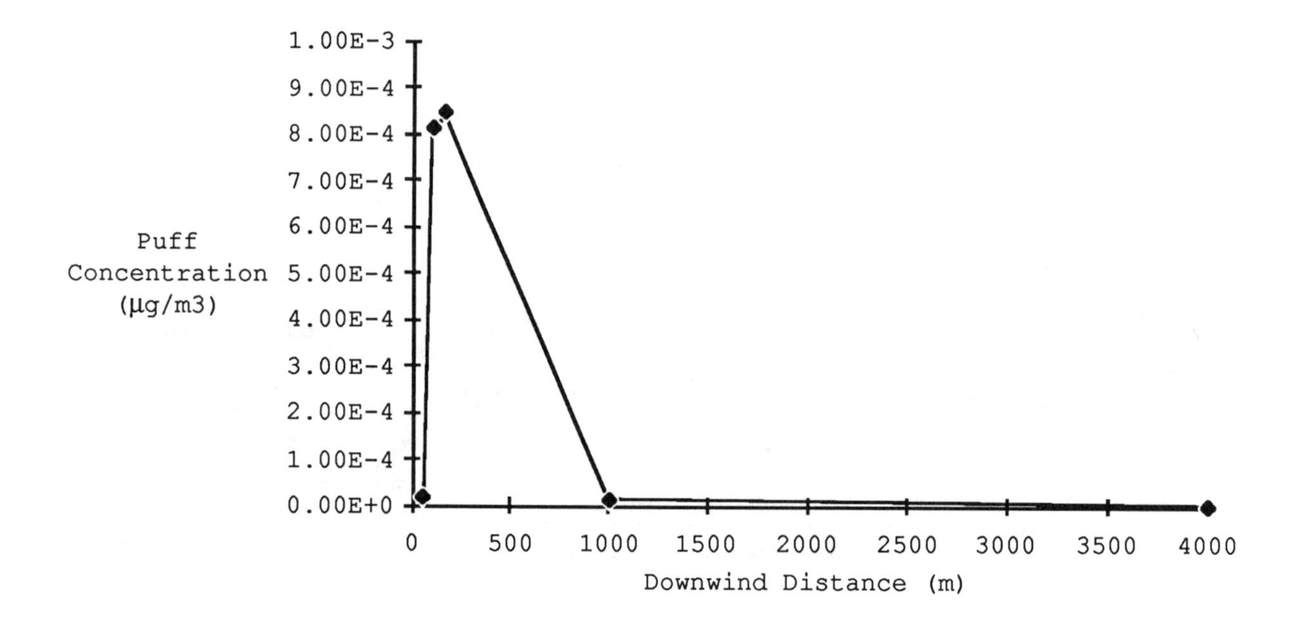

b. The dosage of this uncombusted waste feed at each downwind location is based on a 30 min averaging time. Based on a wind speed of 2 m/s, the distance traveled from the source over the averaging period is:

$$\text{Distance} = u\ (30\ \text{min}) = 2\ \text{m/s}\ (30\ \text{min})\ (60\ \text{s/min}) = \mathbf{3600\ m}$$

This distance represents a variety of plume spreads depending on the location at which the determination is being made as:

$$\text{Number of } \sigma_x \text{ at a given location} = 3600\ \text{m}/\sigma_x \text{ at that location}$$

These results are tabulated below along with the average/peak concentration = 2.5067/m, the averaged concentration, C_{ave}, and the dose = C_{ave} (30 min) in units of μg s/m^3.

RA-5 Solution (Continued).

Distance (m)	σx (m)	m	[Ave.]/[Peak]	Cave	Dose (μg s/m^3)
50	4.2	857.1	0.00292	23.6	708
100	8.0	450	0.00557	1892.3	56,769
160	12.5	288	0.00870	6419.3	192,579
1000	68.0	52.9	0.0473	323.7	9711
4000	245	14.7	0.17	41.6	1248

RA-6 Solution.

a. The metal particulate emission rate is given as:

Q = (0.08 gr/dscf) (1 lb/7000 gr) (10,000 dscf/min) (60 min/h)
= **6.86 lb/h = 0.865 g/s**

b. The settling velocity of the particles is calculated using the terminal velocity equation for the three particle diameters 0.5, 5 and 50 μm as:

$$v_t = \frac{D_p^2\ \rho_p\ g}{18\ \mu} = \frac{D_p^2\ (2.5\ \text{g/cm}^3)\ (980.7\ \text{cm/s}^2)}{18\ (182.7 \times 10^{-6}\ \text{g/cm-s})}$$

$$= \mathbf{D_p^2\ (7.46 \times 10^5)\ ;\ D_p\ \text{in cm}}$$

Using this equation, the following table can be constructed yielding settling velocities and travel distances for a wind speed of 5 mi/h, with travel time = H/v_t = 200 ft/v_t = 61 m/v_t, and travel distance = travel time (5 mi/h):

Particle Diameter (μm)	Settling Velocity (m/s)	Travel Distance (mi)
0.5	1.9×10^{-5}	4459
5.0	1.9×10^{-3}	44.6
50	1.9×10^{-1}	0.446

c. The distance at which the centerline of the tilted plume would touch the ground surface in miles is given in the table above as Travel Distance (mi). Particles with a diameter < 1 μm in diameter settle very slowly. Their residence time in the atmosphere is dictated primarily by wet and dry removal mechanisms, particulate and droplet condensation reactions, etc., not gravitational mechanisms.

d. The gaussian plume equation is valid for relatively short distances, i.e., $x < 50$ km; thus, these ambient concentrations are calculated only for the 50 μm particles where the travel distance is 0.446 mi = 719 m. For this condition:

$$\sigma_y = 0.32\ (719)^{0.78} = \mathbf{54\ m} \qquad \sigma_z = 0.22\ (719)^{0.78} = \mathbf{37\ m}$$

The gaussian plume model for particulates shown below can be used to calculate downwind particulate concentrations.

$$C(x,y,0;H) = \frac{Q}{2\ \pi\ u\ \sigma_y\ \sigma_z}\ \exp\left[-\frac{1}{2}\left(\frac{y}{\sigma_y}\right)^2\right]\ \exp\left[-\frac{1}{2}\left(\frac{H - v_t\ (t)}{\sigma_z}\right)^2\right]$$

RA-6 Solution (Continued).

The $H - v_t(t)$ term, with t = time to a downwind distance x, accounts for the trajectory of the centerline of the particulates in the z direction as they settle. With $H - v_t(t) = 0$ m (effective height of the plume adjusted to ground level at 719 m where the 50 μm particles have reached the ground), y = 0 m (centerline), and u = 5 mi/h = 2.2 m/s, the solution to this gaussian plume equation then becomes:

$$C(719,0,0;61) = \frac{Q}{2\pi u \sigma_y \sigma_z} = \frac{865 \text{ mg/s}}{2\pi (2.2 \text{ m/s})(54 \text{ m})(37 \text{ m})}$$

$$= 0.0313 \text{ mg/m} = \mathbf{31.3\ \mu g/m^3}$$

e. The surface flux (D) due to settling is $D = (C)(v_t)$. At the 0.25 mi = 403 m downwind location, the centerline concentration, y = 0, is calculated at an initial effective stack height H = 61 m, and a travel time of t = 0.25 mi/(5 mi/h) = 0.05 h = 6 min = 180 s. The dispersion coefficients are calculated as:

$$\sigma_y = 0.32\ (403)^{0.78} = \mathbf{34\ m} \qquad \sigma_z = 0.22\ (403)^{0.78} = \mathbf{24\ m}$$

Using these coefficients, the gaussian dispersion equation for the 50 μm particles becomes:

$$C(403,0,0;61) = \frac{865 \text{ mg/s}}{2\pi (2.2 \text{ m/s})(34 \text{ m})(24 \text{ m})} \exp\left[-\frac{1}{2}\left(\frac{61 \text{ m} - 0.19 \text{ m/s } (180 \text{ s})}{24 \text{ m}}\right)^2\right]$$

$$= 0.0767 \text{ mg/m}^3\ (0.536) = \mathbf{41.1\ \mu g/m^3}$$

For the 5 μm particles, the gaussian dispersion equations becomes:

$$C(403,0,0;61) = \frac{865 \text{ mg/s}}{2\pi (2.2 \text{ m/s})(34 \text{ m})(24 \text{ m})} \exp\left[-\frac{1}{2}\left(\frac{61 \text{ m} - 0.0019 \text{ m/s } (180 \text{ s})}{24 \text{ m}}\right)^2\right]$$

$$= 0.0767 \text{ mg/m}^3\ (0.041) = \mathbf{3.14\ \mu g/m^3}$$

For the 0.5 μm particles the settling velocity correction can be ignored, and the gaussian dispersion equations simplifies to:

$$C(403,0,0;61) = \frac{865 \text{ mg/s}}{2\pi (2.2 \text{ m/s})(34 \text{ m})(24 \text{ m})} \exp\left[-\frac{1}{2}\left(\frac{61 \text{ m}}{24 \text{ m}}\right)^2\right]$$

$$= 0.0767 \text{ mg/m}^3\ (0.0396) = \mathbf{3.03\ \mu g/m^3}$$

The deposition rate, D, for each of the particle sizes is thus:

$$D(0.5\ \mu\text{m}) = 3.03\ \mu\text{g/m}^3\ (1.9 \times 10^{-5} \text{ m/s}) = \mathbf{5.8 \times 10^{-5}\ \mu g/m^2\text{-}s}$$
$$D(5.0\ \mu\text{m}) = 3.14\ \mu\text{g/m}^3\ (1.9 \times 10^{-3} \text{ m/s}) = \mathbf{6.0 \times 10^{-3}\ \mu g/m^2\text{-}s}$$
$$D(50\ \mu\text{m}) = 41.1\ \mu\text{g/m}^3\ (1.9 \times 10^{-1} \text{ m/s}) = \mathbf{7.8\ \mu g/m^2\text{-}s}$$

f. The mass of 0.5 μm diameter particles deposited in a 1 m^2 area during a 30 yr period is:

RA-6 Solution (Continued).

Mass 0.5 μm particles = 5.8×10^{-5} μg/m^2-s (1 m^2) (30 yr) (3.1536×10^7 s/yr)
= 54,872.6 μg (0.05) {% time with stated met conditions}
= **2744 μg for this particular modeling condition**

The resultant concentration of these 0.5 μm particles in the upper 10 cm layer of soil is:

C(0.5 μm) = mass of particulates/mass of soil
= 2744 μg/[(1 m^2)(2×10^6 g/m^3)(0.1 m)] = 2744 μg/200,000 g = **0.0137 ppm**

For the 5 and 50 μm particles, similar calculations can be carried out with the following results:

C(5 μm) = **1.3 ppm**
C(50 μm) = **185 ppm**

g. The soil metal concentrations calculated in Part f correspond to those in the problem statement. The annual probability of an individual dying from a carcinogenic effect from metal deposition over a 30 year period is given by the product of the probabilities in subparts 1, 2, 3, and 4 of this question for a particular particle size and corresponding particle soil concentration.

Based on these numbers, the following probabilities of carcinogenic effects from this metal fallout are as follows:

p(0.5 μm) = (0.25)(0.8)(0.0001)(0.005) = **1×10^{-7}**
p(5 μm) = (0.25)(0.8)(0.001)(0.005) = **1×10^{-6}**
p(50 μm) = (0.25)(0.8)(0.1)(0.005) = **1×10^{-4}**

h. Most of the particles emitted from a controlled hazardous waste incinerator will be small, probably near 1 μm diameter. These particles will disperse like a gas, and gravity settling will be relatively unimportant. This analysis shows the sensitivity of predicted health effects to particle size, and indicates the magnitude of potential impacts. Improved treatment of settling and removal processes occurring in the atmosphere, i.e., wet and dry deposition, may greatly refine these predictions, as would the use of simulation models such as ISC-LT or PEM-2 which can more accurately account for long term meterological conditions and particle transport and dispersion.

CHAPTER NINE

Monitoring

M-1 Solution.

The detection limit for hexachlorobenzene is 1 ng/μL = 1 μg/mL = 1 mg/L = 0.001 g/L. If the sample is concentrated in 25 mL solvent, the amount of flue gas to be collected in scf is:

Concentration of hexachlorobenzene in flue gas:

Concentration = 5000 lb/h (454 g/lb) (1 h/60 min)/(427,000 scfm) (0.01) (1-0.9999) = **0.0886 μg/scf**

Sample Volume = 1 μg/mL (25 mL) (1 scf)/0.0886 μg = **282 scf**

If the sample is collected at 1 L/min, then the time for sampling is:

$$\text{Time} = (282\ \text{scf})\ (28.3\ \text{L/ft}^3)\ (1\ \text{min/L}) = \mathbf{7985\ min = 133\ h}$$

M-2 Solution.

a. Minimum concentrations for detection in the stack gas for a 20 L sample are calculated as follows:

$$CCl_4\text{: } 200\ \text{ng in } 20\ \text{L} = 10\ \text{ng/L} = 10\ \mu\text{g/m}^3$$
$$\text{TCE: } 250\ \text{ng in } 20\ \text{L} = 12.5\ \text{ng/L} = 12.5\ \mu\text{g/m}^3$$
$$\text{DCM: } 500\ \text{ng in } 20\ \text{L} = 25\ \text{ng/L} = 25\ \mu\text{g/m}^3$$

With a stack gas flow rate = 500 dscm/min, the mass flow out is as follows:

$$(w_{out})CCl_4 = (10\ \mu\text{g/m}^3)(500\ \text{dscm/min}) = 5000\ \mu\text{g/min} = 5\ \text{mg/min}$$
$$(w_{out})TCE = (12.5\ \mu\text{g/m}^3)(500\ \text{dscm/min}) = 6250\ \mu\text{g/min} = 6.25\ \text{mg/min}$$
$$(w_{out})DCM = (25\ \mu\text{g/m}^3)(500\ \text{dscm/min}) = 12500\ \mu\text{g/min} = 12.5\ \text{mg/min}$$

With a DRE of 99.99%, the remaining mass of POHC is (1 - 0.9999) = 0.0001, therefore, the mass flow into the incinerator must be as follows:

$$(w_{in})CCl4 = (w_{out})CCl_4/0.0001 = 5\ \text{mg/min}/0.0001 = 50\ \text{g/min} = 3\ \text{kg/h}$$
$$(w_{in})TCE = (w_{out})TCE/0.0001 = 6.25\ \text{mg/min}/0.0001 = 62.5\ \text{g/min} = 3.75\ \text{kg/h}$$
$$(w_{in})DCM = (w_{out})DCM/0.0001 = 12.5\ \text{mg/min}/0.0001 = 125\ \text{g/min} = 7.5\ \text{kg/h}$$

For each POHC, a safety factor of 10 is to be used to insure detectability; therefore, each final inlet rate should be increased by 10 to yield:

$$(win)CCl_4 = 30\ \text{kg/h}$$
$$(win)TCE = 37.5\ \text{kg/h}$$
$$(win)DCM = 75\ \text{kg/h}$$

b. The stack gas concentration is recalculated for a DRE of 99.98% to determine a revised mass outflow rate:

$$(w_{out})CCl_4 = (0.0002)(30\ \text{kg/h}) = 6\ \text{g/h} = 100\ \text{g/min}$$
$$(w_{out})TCE = (0.0002)(37.5\ \text{kg/h}) = 7.5\ \text{g/h} = 125\ \text{g/min}$$
$$(w_{out})DCM = (0.0002)(75\ \text{kg/h}) = 15\ \text{g/h} = 250\ \text{g/min}$$

The flue gas concentration in 500 dscm/min is calculated as follows:

$$(w_{out})CCl4 = (100\ \text{mg/min})/(500\ \text{dscm/min}) = 200\ \mu\text{g/m}^3 = 200\ \text{ng/L}$$
$$(w_{out})TCE = (125\ \text{mg/min})/(500\ \text{dscm/min}) = 250\ \mu\text{g/m}^3 = 250\ \text{ng/L}$$
$$(w_{out})DCM = (250\ \text{mg/min})/(500\ \text{dscm/min}) = 500\ \mu\text{g/m}^3 = 500\ \text{ng/L}$$

The calculated loadings on Tenax™ for 20 L samples are as follows:

M-2 Solution (Continued).

CCl_4 Loading = (200 ng/L) (20 L) = 4000 ng, at maximum limit
TCE Loading = (250 ng/L) (20 L) = 5000 ng, at maximum limit
DCM Loading = (500 ng/L) (20 L) = 10,000 ng, at 2 times maximum limit

Since the DCM loading is two times the maximum recommended recovery limit, breakthrough for the DCM will be unacceptable. The loadings for CCl_4 and TCE are at the maximum loading levels and therefore breakthrough will also be a concern for these POHCs.

M-3 Solution.

a. Looking at the boiling point data for the compounds listed and comparing these values to the criteria of b.p. ≥ 100°C for VOST and 30 to 100°C for semi-VOST, the following distribution can be given:

VOST	Semi-VOST
carbon tetrachloride	hexachlorobenzene
1,1,1-trichloroethane	heptachlor epoxide
tetrachloroethylene	chlordane
	decachlorobiphenyl
	pentachlorophenol
	kepone
	hexachlorobutadiene

b. The distribution will not be all or nothing. Some of the less volatile compounds from the VOST procedure will be collected in the Semi-VOST system and vice versa. Compounds will have substantial vapor pressures below their boiling points so the dividing line between analytical methods is not distinct.

c. If the true value of the DRE = 0.9999, then the true value of the mass flow out of the incinerator is:

$$w_{out} = (0.0001)\ (1.0\ kg/h) = 0.1\ g/h$$

If the analytical tolerance is the true value ± 50%, then the range of the mass flow out of the incinerator will be 0.05 g/h to 0.15 g/h. Because the DRE is defined as:

$$DRE = \left(1 - \frac{w_{out}}{w_{in}}\right) 100$$

then the allowable measured range of the true DRE will be 99.985% to 99.995% for a true DRE of 99.99%.

d. Heat of combustion values for these compounds provides the following relative combustion difficulty ranking:

		kcal/g
Supposedly Most Difficult to Incinerate	carbon tetrachloride	0.24
	tetrachloroethylene	1.19
	hexachlorobenzene	1.79
	1,1,1 trichloroethane	1.99
	pentachlorophenol	2.09
	hexachlorobutadiene	2.12
	kepone	2.15
Supposedly Least Difficult to Incinerate	decachlorobiphenyl	2.31
	heptachlor epoxide	2.70
	chlordane	2.71

M-3 Solution (Continued).

e. A thermodynamic property may be appropriate for predicting equilibrium conditions, but the rapid reactions in an incinerator will be governed by kinetic considerations as well. Consequently, a measure of incinerability based on reaction energy and destruction kinetics would be a more relevant indicator of the difficulty of combustion for a given chemical.

PART 3

USER'S GUIDE FOR THE *HWI SOFTWARE*

The purpose of the *User's Guide* is to provide all necessary software documentation and instructions for using the three programs contained on the enclosed computer diskette. A companion *Reference Manual* (see Part 4 of this book) is intended to familiarize the user with the basic scentific and engineering principles behind incineration calculations.

CHAPTER ONE

Introduction to the *HWI* Software Package

<u>What is the HWI Software Package?</u>

The HWI software package is contained on the floppy disk that accompanies this Guide. This package consists of three programs and two data files. The programs, which are discussed in detail in Chapters 3, 4 and 5 of this Guide, are named *HWI*, *HWISET* and *HWITRL*; the executable code for each of these programs resides in a file of the same name with the extension *.EXE* (i.e., the machine language version of *HWI* is a file named *HWI.EXE*, etc.). The accompanying data files are named *HWI.DAT* and *HWISET.DAT*. A brief description of each of the programs appears below:

1. *<u>HWI</u>*. This program performs the hazardous waste incinerator calculations. Communication between you and the program is handled in interactive fashion; all data are inputted in response to prompts that appear on the monitor screen during execution of the program. Both data files *HWI.DAT* and *HWISET.DAT* supply required information to *HWI*.

2. *<u>HWISET</u>*. This is a utility program for use with *HWI*. Its purpose is to allow you more efficient use of the *HWI* program by placing in your hands some control over which input prompts will appear on the screen, which results will be displayed, and which calculations can be omitted during execution of the *HWI* program. The choice of alternate input options may be made beforehand, relieving the user of the task of choosing these options while the *HWI* program is running. *HWISET* generates the data file *HWISET.DAT*, which is read by *HWI* at the start of its run.

3. *<u>HWITRL</u>*. The "*TRL*" stands for tutorial. The purpose of *HWITRL* is to familiarize the user with the *HWI* program. The solutions to two problems, as they would be solved using *HWI*, are simulated in step-by-step fashion. (These are the same two problems that are solved in detail in Parts IV and V of the *Reference Manual*). Some explanation of the theory behind the calculations is also presented on the monitor during the execution of *HWITRL*.

What Type of Computer is Required to Run HWI?

For the HWI software package to run on your system, your computer must be an IBM or IBM-compatible of the XT class or higher. Although a hard-disk system is preferable (for speed in executing the program), the package will also run on a floppy-disk system. The monitor may be either monochrome or color. A Disk Operating System such as PC-DOS or MS-DOS must be available on either the hard disk (if a hard-disk system is being used) or a floppy diskette.

Directions for using the HWI package with several hardware configurations are given in Chapter 2.

CHAPTER TWO

Preparing to Use *HWI*

In this chapter you will find detailed instructions for installing and running the *HWI* software program. Installation instructions are provided for both hard disk and floppy disk users.

The *HWI* program requires PC DOS or MS DOS in order to run. Due to licensing restrictions, the diskette supplied does not contain the necessary DOS system files on it. However, by using a few DOS commands, you can create a working copy of your *HWI* PROGRAM DISK. Your working copy will be used to run the program; you should store the original diskette in a safe place. This way you will be protected in the event that the working copy becomes damaged.

The simple procedures for making your working copy are described in the following sections. Since there are differences in the way that hard disk-based system users (e.g., PC XT) and diskette-based system users (e.g., PC) will set up working software, these two systems are described separately. Follow the instructions that apply to your system.

Making a Working Copy on Diskette

To create a working copy of the *HWI* program on diskette, you will need your DOS master diskette, the *HWI* diskette and either a blank diskette or a diskette which contains information that you no longer need. Place the DOS diskette in drive A and turn on your PC. If the computer is already on, hold down the two keys marked **Ctrl** and **Alt** and press the key marked **Del**. After you enter the date and time, the display will show the DOS prompt, A:>, and a flashing cursor which indicates that DOS is ready to accept commands.

You must first format your blank diskette and put you version of the operating system on it. Keep your DOS diskette in drive A and place the blank diskette in drive B. Type in the following command after the A:> prompt:

FORMAT B:/S (then press ENTER)

Follow the prompts on your screen from the FORMAT command. If you have only one floppy disk drive you can still use the same command; DOS will treat that drive as both A and B and will tell you when to switch diskettes. When FORMAT is finished, your once blank diskette will have all except two of the necessary DOS system files on it.

These additional DOS files, ANSI.SYS and CONFIG.SYS should be copied from the DOS diskette onto your working diskette. Just use the DOS copy command, as shown below:

COPY A:ANSI.SYS B: (then press ENTER)

COPY A:CONFIG.SYS B: (then press ENTER)

Now you can transfer the actual *HWI* files to the new diskette. Remove the DOS master diskette and place the original *HWI* PROGRAM DISK in drive A and use the COPY command one more time:

COPY A:*.* B: (then press ENTER)

Once all the files have been copied, your working diskette will be ready to use. The original diskette should be stored in a safe place and you should use the working copy from now on. Please continue with the section "CHECKING YOUR CONFIG.SYS FILE".

Making a Working Copy on a Hard Disk

PC XT, PC AT, or other hard disk users may choose to put the files from the *HWI* disk on the hard disk so it will not be necessary to maintain a separate diskette. Of course, you may elect not to transfer the software to the hard disk. In that case, you should follow the directions in the previous section to make a working diskette copy.

The instructions that follow assume that you have already prepared your hard disk according to chapter titled "Preparing Your Fixed Disk" in your DOS manual. It is also recommended that you become familiar with a few DOS commands which pertain to the usage of subdirectories. Please refer to the chapter "Using Tree-Structured Directories" in your DOS manual.

Start up your system from the hard disk and place your *HWI* diskette in drive A. Since this diskette consists of several files, the best thing to do is keep the files on the hard disk in their own subdirectory. One method of doing this is described below. Users who have their own, specific file organization scheme for their hard disk may choose to apply another method.

Before you use a new subdirectory, you must create it. You can make up your own name for it, but we'll use \HWI in this example. After the DOS prompt, C:>, enter the following:

MD \HWI (then press ENTER)

3-6

That will create the subdirectory for you. Once it's created, make it the current directory by entering this command:

CD \HWI (then press ENTER)

Now copy all the files on the *HWI* Program Disk to the hard disk with this command:

COPY A:*.* (then press ENTER)

Your \HWI subdirectory is now completely set up. Please go on to the next section for information about your CONFIG.SYS file.

CHECKING YOUR CONFIG.SYS FILE

Before you start the *HWI* program please check to see that your CONFIG.SYS file has the following lines:

```
FILES=10          (or more)
BUFFERS=10        (or more)
DEVICE= ANSI.SYS
```

If you are using a floppy disk-based system, your working copy of the software is probably still in drive B, if so, type:

TYPE B:CONFIG.SYS (then press ENTER)

Hard disk users will find their CONFIG.SYS file in their root directory, type **CD ** to make the root directory the current directory and issue the command:

TYPE CONFIG.SYS (then press ENTER)

The contents of the CONFIG.SYS file should appear on the screen. Check the make sure that the above three lines are in the file. The order that these lines appear in isn't important, but all three lines must be in the CONFIG.SYS file. If your FILES= and/or your BUFFERS= lines show more than 10 (e.g., 35) that is fine; 10 is the minimum value for these settings. The DEVICE=ANSI.SYS line is especially important, since this command line determines how your video display will look.

If your CONFIG.SYS file does not contain any/all of the above lines, it is necessary to modify it. Please refer to your DOS manual to the chapter "Changing the System Configuration" for instructions on how to modify your CONFIG.SYS file.

Keep in mind that if you need to modify the CONFIG.SYS file, it is necessary to reboot (restart) the computer for the changes to take effect.

STARTING THE PROGRAM

If you copied the *HWI* program to a hard disk, skip to the next paragraph to see how to start up. Otherwise, place your working diskette copy in drive A and start your system as described in the beginning of the section "Making a Working Copy on Diskette." Or, if you already have a DOS prompt on your screen, enter the desired *HWI* program name (**HWI**, **HWISET** or **HWITRL**) to start the program. For example:

HWI (press the ENTER key)

Users who copied the program to a hard disk using the directions under "Making a Working Copy on a Hard Disk" can access the program by using this method:

1. Start your system (if it's not already on) from the hard disk.

2. Enter **CD \HWI** after the C:> prompt to make the HWI directory the current directory.

3. Enter **HWI**, or **HWISET**, or **HWITRL**, press the ENTER key, to start up the desired program.

If you are familiar with DOS batch commands, you can create your own batch command file to issue these command sequence for you.

When you start the program, you should be greeted with a logo display followed by a title/copyright screen. When you press a key, as the message indicates, you will see the initial screen of the program you selected, and you can continue with the program input as described in the following chapters.

CHAPTER THREE

The Program *HWITRL*

How to Run *HWITRL*

The program *HWITRL* is accessed in the same manner that all of the programs are accessed, that is, simply by typing the name of the program at the prompt (usually >C or >A), followed by depressing the **RETURN** or **ENTER** key. After a few seconds, the opening frame will appear on the monitor.

Since *HWITRL* is self-explanatory (i.e., instructions for its use appear on the screen while the program is running), a detailed description of it will not be given here.

What *HWITRL* Does

The purpose of *HWITRL* ("*TRL*" stands for "tutorial") is to familiarize you with the program *HWI*. The tutorial program does not actually perform any calculations the way *HWI* does, but it does simulate these calculations, and the results that appear on the monitor screen are the same ones that would be generated by *HWI*. During the execution of *HWITRL*, two problems will be presented and then solved. (These two problems are listed at the end of this chapter and are solved in detail in the *Reference Manual*.) In the first, you are lead in step-by-step fashion through the data input stage and then the calculation stage. During the input phase, you are requested to answer the prompts in much the same way that the input prompts are answered in the *HWI* program. One difference is that, in *HWITRL*, the answer to each prompt is checked by the program to make sure it is the correct one. If it is not, the computer responds with a rather irritating noise, and you are again requested to answer the prompt. This process continues until the prompt is answered correctly. This, of course, is not done when *HWI* is being run. Unlike the tutorial, the *HWI* program does not "know" whether or not you have inputted correct data.

During the calculational phase of the *HWITRL* run, the calculations that are "going on" are explained on the monitor screen and many of the intermediate results that would not be displayed by *HWI* are shown. Once again it should be pointed out that during the running of *HWITRL*, the "calculations" are not really being performed, but merely simulated.

After observing the "solving" of Problem 1, the second problem is then presented on the screen. This time, the input prompts are not as generous as before. A portion of the problem statement appears before each prompt and you are expected to extract the necessary information from the statement to answer the prompt. Again, if an error is made, *HWITRL* will let you know and give you a chance to answer the prompt correctly.

Correcting Input Errors

In order to enter data to *HWITRL* (or, for that matter, to any of the three programs), the typing of the data at the keyboard must be followed by depressing the **RETURN** or **ENTER** key. The data will not be "read" by the program until this key is depressed. This feature gives you a chance to correct the input if you have made an error. To correct or change an input that has just been typed but has not yet been "entered" by depressing the **RETURN** or **ENTER** key, simply hit the **BACKSPACE** key one or more times until the input (or the incorrect portion of it) has been erased. Then type the new input.

If the **RETURN** or **ENTER** key has been depressed before you notice the error, *HWITRL* lets you know immediately and gives you a chance to correct it by repeating the prompt. In fact, *HWITRL* will not let you proceed until you enter the data correctly. (Note: If you enter incorrect data during the running of *HWI*, an error message will be indicated only if it is an obvious error, for example, a letter or other invalid character appearing in what should be a numerical input.)

Fast-Forward, Reverse and Stop

HWITRL is a fairly long program and you may not want to go through the whole program at one sitting. You may also, at times, wish to go back to a previous section to review material presented earlier. The directions for terminating, moving forward and moving backward through the program are given below.

1. Stopping the program. If you wish to terminate the program at any time, simply depress the **CRTL** (Control) and **C** keys simultaneously and the DOS prompt (usually **C>** or **A>**) will appear.

2. Reverse. If you wish to "back up" to some previous point in the program, input a "-1", "-2", or "-3" whenever the following prompt is displayed by *HWITRL*:

   ***** Depress the RETURN or ENTER key to continue *****

   This message comes up many times during the execution of *HWITRL*. Inputting a "-1" will cause the program to return to the beginning of the present section. (A "section" in *HWITRL* generally, but not always, begins with a heading that is identified by a capital letter, e.g., *"A. Input of Waste Composition and Flow Rate"; "B. Calculation of Stoichiometric Air and Input of Percent*

Excess Air"; etc.). Inputting a "-2" returns the program to the beginning of the previous section and "-3", to the beginning of the section before that. For example, suppose the frame with the heading: "*E. Calculation of Incinerator Temperature and Flue Gas Composition*", appears on the screen. Inputting a "-1" at this point simply brings up the same frame again, since this frame is the start of the current section. Inputting a "-2" brings up the frame with the heading: "*D. Input of Preliminary Design Parameters*", which is the start of the previous section.

3. Fast Forward. If you wish to "jump ahead" in the program, as would be the case if you are going through the tutorial in more than one sitting and wish to skip over material already covered, input a "1", "2" or "3" in answer to the prompt:

***** Depress the RETURN or ENTER key to continue *****

The "1" moves the program ahead to the beginning of the next section; "2" to the beginning of the section after that; and "3", to the section after that one. Once again, using the frame with the title: "*E. Calculation of Incinerator Temperature and Flue Gas Composition*" as an example, inputting a "1" brings up the beginning of the next section, the lead frame of which is entitled by: "*F. Calculation of the Flue Gas Flow Rate*".

The Illustrative Problems Solved in *HWITRL*

The two problems that are presented and "solved" by *HWITRL* are listed below. Detailed solutions to both of these problems are given in Chapters 4 (Problem 1) and 5 (Problem 2) of Part 4 of this book.

Problem 1:

5000 lb/hr of a hazardous waste is to be incinerated in a liquid injection unit with 100% excess air. The waste has the following composition:

Element	Mass %
carbon	60.8
hydrogen	4.2
chlorine	30.0
sulfur	5.0

a. Calculate the flue gas composition and flow rate assuming complete combustion.

b. Calculate the incinerator temperature.

c. Determine the length and diameter of the incinerator. Assume a flue gas superficial throughput velocity of 20 ft/sec.

Problem 2:

A hazardous waste has the following composition:

Component	Mass %
Chlorobenzene, C_6H_5Cl	58
DDT, $C_{14}H_9Cl_5$	26
Water, H_2O	16

Calculate the flue gas flow rate and composition when 6430 lb/hr of the waste is incinerated in a rotary kiln incinerator with 75% excess air. Also calculate the operating temperature and dimensions of the cylindrical housing. Assume an overall L/D ratio of 4.0 for the kiln/afterburner unit.

From heat of formation data, the following values for the heat combustion (NHV) of the three components can be calculated. These are:

Chlorobenzene	714,361 cal/gmol
DDT	1,580,000 cal/gmol
Water	-10,519 cal/gmol

(The "heat of combustion" listed for water is actually the heat of vaporization. In the incinerator, the water simply vaporizes.)

CHAPTER FOUR

The Program *HWI*

What *HWI* Does

HWI is the main program of the HWI software package and handles all of the calculations involved in the simulation of a hazardous waste incinerator. These calculations may be divided into three main categories:

(1) thermochemical calculations, which relate operating incinerator temperature, excess air, and feed (including not only waste but also possible fuel requirements) heating value;

(2) stoichiometric calculations, which yield the composition and flow rates of gaseous emissions from incinerators burning (hazardous) waste; and

(3) preliminary incinerator design.

It should be noted that the sequence listed above is not the sequence in which these calculations are actually performed in the *HWI* program. The calculations of Categories (1) and (2), for example, are interdependent and must be treated in simultaneous fashion (cf. Chapter 3 of the *Reference Manual*, Part 4 of this book).

For What Type of Waste Is *HWI* Intended?

The waste (or waste-fuel mixture, when auxiliary fuel has been added to enhance the heating value) must be primarily organic and may be liquid, solid, or a combination (sludge, slurry, etc.). It may contain any or all of the following elements: carbon, hydrogen, oxygen, nitrogen, sulfur, chlorine, fluorine, bromine, and iodine. *HWI* will not recognize any other elements besides these. Inert solid material (designated in this *Guide* by the term "**ash**") may be included, but the amount should not exceed about 5% on a mass basis. It is assumed during the calculations that any ash present in the waste has negligible impact on incinerator performance and all of it leaves the incinerator as particulate matter entrained in the flue gas stream.

How to Run *HWI*

To access *HWI*, type the letters **HWI** at the DOS prompt (usually **C>** or **A>**) and depress the **RETURN** or **ENTER** key. After a few seconds, the title frame will appear on the monitor.

Inputting information to *HWI* is accomplished at the computer terminal and is handled in question-and-answer fashion. The question or request for a particular input appears on the screen and the user types in the answer. Each typed input must be entered by depressing the **RETURN** or **ENTER** key. Until this is done the program will not "read" the information that was typed.

Most of the inputs are *numerical*; a few (for example, chemical symbols) are *alphabetical*. Besides numerical digits, a *numerical* input allows only for a decimal point, a plus (+) sign or a minus (-) sign. The plus or minus sign must appear as the first character of the number. Minus signs are mandatory for negative numbers; plus signs are optional for positive numbers. Decimal points are also optional if the number is an integer. **516.0** may simply be input as **516**, for example. Any other non-numerical character will cause an error message to come up on the monitor screen. The number **71,600**, for instance, must be typed without the comma, i.e., as **71600**. Units such as **%** or **lb/hr** are never entered. If you type a number with an illegal character in it, *HWI* will respond by repeating the input prompt, thereby giving you a chance to type the number correctly. Note that if you type and enter an incorrect number, and this number is accepted by the program because *HWI* cannot find anything wrong with it, you still have a chance to correct it by using the "backtrack" feature, the details of which will be described in the next section (*How to Correct Input Errors to HWI*).

The only *alphabetical* inputs to *HWI* are the chemical symbols of the elements. The symbols "recognized" by *HWI* are:

C for carbon
H for hydrogen
O for oxygen
Cl for chlorine
S for sulfur
N for nitogen
F for fluorine
Br for bromine
I for iodine
WA for water
AS for ash

The last two are not standard element symbols; their use will be explained later. When any of these eleven symbols is typed as an input to *HWI*, it must be typed exactly as it appears above. For example, if **CL** is typed for chlorine instead of **Cl**, an error message will appear. In this case, the prompt will also be repeated, giving you a chance to correct the error.

The Novice and Expert Modes

After the opening (title) frame of *HWI*, the next frame to appear on the screen contains one of two possible messages, the first of which is shown in ***bold italics*** below. (Note: In this and the next chapter, ***bold italics*** will be used to distinguish material that appears on the screen during the running of *HWI* from explanatory text.)

This program may be run in one of two modes. The NOVICE mode is intended for the beginner who is not yet familiar with the use of the program. In this mode, the program provides a complete set of prompts that will facilitate the inputting of data. For the more experienced user, the EXPERT mode allows the omission of many of the prompts and permits the choosing of several input options to be made beforehand. The program is presently set to the NOVICE mode. The User's Guide should be consulted for directions on the use of the EXPERT mode.

The next chapter of this *Guide*, Chapter 5, describes the use of the NOVICE and EXPERT modes in detail. If the above message is the one to appear, the *HWI* program is obviously set to the NOVICE mode and the reader who has little or no experience in running *HWI* (and who shall hereafter be referred to in this *Guide* as the "novice reader") should ignore the remainder of this section and continue on to the next section (*Inputting Data to HWI*). If the program is set to the EXPERT mode, the following message is the one to appear.

This program is presently set to run in the EXPERT mode. To change to the NOVICE mode, depress the CTRL and C keys simultaneously. (This will cause the program run to abort.) Then type "HWISET", depress the RETURN or ENTER key, and follow the instructions that appear on the screen.

If the program is, as indicated by the above message, set to the EXPERT mode, the NOVICE reader should follow the instructions to abort the program (i.e., depress the **CTRL** and **C** keys simultaneously), and type the letters **HWISET** in response to the DOS prompt. When the **RETURN** or **ENTER** key is then depressed, the title frame for the *HWISET* program appears. If the **RETURN** or **ENTER** key is depressed five times more, *HWISET* will terminate. It is not necessary at this point to read the material that appears on the screen. When the DOS prompt appears again, the letters **HWI** should be typed to restart the *HWI* program. This time the message on the screen will indicate that *HWI* is set to the NOVICE mode.

Inputting Data to *HWI*

The required data to be inputted by the user during the running of *HWI* include: calculational mode, waste composition, waste flow rate, air requirements, waste heating value, heat loss from the incinerator, and some design information. Each of these is described in detail below.

1. Calculational mode. The three key parameters in the *HWI* calculations are the incinerator operating temperature, the waste-fuel heating value and the amount of excess air. The waste-fuel heating value is determined by the materials being combusted. Either one of the remaining two parameters may be treated as the independent variable. The *HWI* program provides two options: (1) the amount of excess air to be employed may be determined by the user, and the resulting incinerator operating temperature calculated by *HWI*; or (2) the desired operating temperature may be inputted by the user and the required amount of excess air calculated by the program. These two modes are described in the next frame (Main Menu frame) to appear on the monitor screen, as shown below.

M A I N   M E N U

The following options are available in HWI:

---

1 - The waste (or waste-fuel mixture) composition and flow rate, plus the percent of excess air are to be specified. The incinerator operating temperature is to be calculated by the program.

2 - The waste (or waste-fuel mixture) composition and flow rate, plus the incinerator operating temperature are to be specified. The percent of excess air required to achieve this temperature is to be calculated by the program.

Input the number corresponding to the option to be used: ...

If any character is entered other than a "1" or "2", or if the **RETURN** or **ENTER** key is depressed without typing anything (This is equivalent to entering zero), *HWI* responds with an error message and repeats the prompt.

2. Waste composition. In order to accommodate various methods of characterizing the chemical make-up of the waste, *HWI* provides the user with the following options:

1 - elemental (or ultimate analysis), atom basis
2 - elemental analysis, mass basis
3 - componential analysis, mole basis
4 - componential analysis, mass basis

In a *componential* analysis, the composition of the waste (or, when auxiliary fuel has been added, the waste-fuel mixture) is described in terms of what chemical *compounds* are present and the fraction or percent of each. In an *elemental* analysis, the composition is given in terms of what *elements* are present and the fraction or percent of each. Whichever of these four options is employed by the user during the input phase of the program, the composition is converted to an elemental analysis on a mole fraction basis before the stoichiometric calculations are begun.

The input menu for the composition of the waste (or waste-fuel mixture) next appears on the screen.

WASTE-FUEL COMPOSITION: INPUT MENU

The composition of the waste-fuel mixture may be inputted in any one of the following four ways:

1 - elemental composition, atom basis *
2 - elemental composition, mass basis
3 - componential composition, mole basis *
4 - componential composition, mass basis

****An atom or mole basis may NOT be used when ash is included in the waste.***

Type the number corresponding to the method to be used for inputting the waste-fuel composition ....

Regarding the asterisked note near the bottom of the frame, the word "ash" refers to material in the waste that is presumed to be inert and have negligible effect on the incinerator performance. (This latter assumption may not necessarily be true if the waste contains a large amount of ash. An amount less than five mass percent of ash is considered reasonable in light of this assumption.) Since the exact composition of the ash is generally unknown, an average molecular weight for it cannot be calculated. When the waste contains ash, therefore, it is not possible to express the waste composition on either an atom (for the elemental composition) or mole (for the componential composition) basis; a mass basis must be used.

The next set of prompts to appear will vary depending on which of the four composition input options is chosen. If one of the elemental analysis options is chosen (Option 1 or 2), the next prompt requests the input of the number of elements in the waste. For each element, the user is then requested to input the element's chemical symbol and its fractional or percentage contribution to the waste (*atom* percent or fraction for Option 1, *mass* percent or fraction for Option 2).

In the following example, the composition of a waste (chlorobenzene) consisting of 64.0 % carbon (by mass), 4.4 % hydrogen and 31.6 % chlorine is inputted using Option 2.

Elemental Input (mass basis):

Note: This program will accept the following element symbols: C, H, O, S, N, Cl, F, Br and I. Water and ash may be treated as "elements", if desired, by using the symbols: WA and AS, respectively. AS will not be accepted when the composition is expressed on a mole or atom basis. Refer to the User's Guide for further information.

Input the number of elements in the waste-fuel mixture: ... 3

For ELEMENT 1:

Input chemical symbol: C

Input mass percent or fraction: 64

For ELEMENT 2:

Input chemical symbol: .... H

Input mass percent or fraction: 4.4

For ELEMENT 3:

Input chemical symbol: Cl

Input mass percent or fraction: 31.6

Although percentages were used to input the waste composition in the example above, fractional or actual amounts may also be used in the *HWI* program. (In fact, any set of numbers that are in the proper ratio will work. For example, instead of 64, 4.4, and 31.6, the numbers 0.640, 0.044, and 0.316 would have the same effect; so would the numbers 32, 2.2 and 15.8).

If the user attempts to input either a zero for any of the numerical inputs above, or an element symbol unrecognizable to *HWI* for any of the alphabetical inputs, the input will be ignored and the prompt repeated. This will also happen if the user tries to input the element symbol "AS" when an atom percent or fraction (Composition Input Option 1) is being used. Again, because ash has no "calculable" atomic weight, its atomic contribution to the waste mixture is indeterminable.

In most elemental analyses, the fraction or percent of any water present in the waste is given as a separate contribution, apart from any oxygen or hydrogen in the waste. In this case, the oxygen and hydrogen in the water cannot be included as part of the oxygen and hydrogen contributions to the waste. (Otherwise, those contributions would be counted twice.) To input the water fraction or percent, the symbol "WA" is used in *HWI*. If Option 1 (atom basis) is being used and water is inputted in this manner, the fact that "WA" consists of three atoms must be taken into account when inputting the atom percent of each component. For example, if a waste consists of equal molar amounts of benzene (C_6H_6) and water, the elemental make-up of the waste may be inputted in either of the following ways:

element	N*	atom %
C	6	40.00
H	8	53.33
O	1	6.67
total	15	100.00

element	N*	atom %
C	6	40.00
H	6	40.00
WA	3	20.00
total	15	100.00

* N = number of atoms in the waste per molecule of benzene.

It should be noted here that if Dulong's approximate method (cf. Chapter 2 of the *Reference Manual*, Part 4 of this book) is to be used to estimate the net heating value of the waste, any water present should be entered as **WA** rather than included in the hydrogen and oxygen in the rest of the waste. This is explained later when the inputting of the net heating value is discussed.

If one of the componential analysis options is chosen (Option 3 or 4), a prompt appears for the number of compounds. For each compound, the user is requested to input the percent or fraction of that compound (*mole* or *mass* percent for Option 3 or 4, respectively) and the number of elements in that compound. For each element in the compound, the user is asked for the chemical symbol and the number of atoms per molecule. Thus, in effect, the user types in the compound's chemical formula. In the next example, the composition of a chlorobenzene-water mixture (80-20 by mass) is inputted using Option 4.

Componential Input (mass basis):

Note: This program will accept the following element symbols: C, H, O, S, N. Cl, F, Br and I. Water or ash may be inputted by using the symbol WA or AS, respectively. In either case, the water or ash is treated as a single component (compound) and is considered to consist of one "element" (viz., WA or AS). AS will not be accepted when the composition is expressed on a mole basis. Refer to the User's Guide for further information.

Input the number of components (different compounds) in the waste-fuel mixture: 2

COMPONENT NUMBER 1:

Input the mass percent or fraction of Component 1: 80

Input the number of elements in Component 1: 3

For ELEMENT 1 of COMPONENT 1:

Input chemical symbol: C

Input number of atoms: 6

For ELEMENT 2 of COMPONENT 1:

Input chemical symbol: .... H

Input number of atoms: 5

For ELEMENT 3 of COMPONENT 1:

Input chemical symbol: C1

Input number of atoms: 1

COMPONENT NUMBER 2:

Input the mass percent or fraction of Component 2: 20

Input the number of elements in Component 2: 2

For ELEMENT 1 of COMPONENT 2:

Input chemical symbol: .... H

Input number of atoms: 2

For ELEMENT 2 of COMPONENT 2:

Input chemical symbol: .... O

Input number of atoms: 1

If the waste contains only one component, the mass (or mole) percent is obviously 100 and the prompt for mass (or mole) percent or fraction does not appear.

As was the case with the elemental composition input options, a zero for any numerical input above or an element symbol unrecognizable to *HWI* will not be accepted by the program. Likewise, the symbol "AS" will not be accepted when a mole percent or fraction (Composition Input Option 3) is being used. If either **WA** or **AS** is entered at a "chemical symbol" input prompt, the prompt for the number of atoms does not appear. (As noted earlier for the elemental composition input, if the net heating value is to be estimated by Dulong's method, any water present should be inputted as WA instead of as H_2O. The reason for this is given later.)

3. Waste flow rate. The waste (or waste-fuel mixture) flow rate is the next input requested. With Composition Option 3 or 4, the user has the choice of *lb/hr* or *lbmol/hr* as shown in the following frame:

Waste-Fuel Flow Rate:

The flow rate of the waste-fuel mixture may be inputted either as lb/hr or lbmol/hr. Type:

"1" if lb/hr is to be used.

"2" if lbmol/hr is to be used.

For Option 1 or 2, *lb/hr* must be used. The next prompt calls for the flow rate; in the example below, the mass flow rate in *lb/hr* is to be inputted.

Input the mass flow rate of the waste-fuel mixture in lb/hr: ...

At this point in the program, a table (or two tables) showing the waste composition appears on the screen. If one of the elemental composition options (Option 1 or 2) was used, a single table for the elemental make-up of the waste appears. If one of the componential composition options (Option 3 or 4) was used, a table for the componential make-up appears first, followed by the elemental table. If the user finds from the table(s) that an error was made in inputting the waste composition, the program may be returned to a previous point (using the backtrack feature to be explained later) and the composition input repeated.

4. Air requirements. If Calculational Mode 1 is being used (i.e., the user inputs the amount of excess air and *HWI* is to calculate the operating temperature), the percent of excess air is inputted at this point. Since the amount of excess air chosen by the user may depend on the amount of stoichiometric or theoretical air (i.e., the exact amount of air required to convert each element in the waste to its combustion product, assuming complete combustion), this quantity is calculated by *HWI* and displayed on the screen prior to the prompt for the percent of excess air. This is demonstrated in the example below.

Air Requirements:

The stoichiometric air requirement is 8.312 lb/lb of waste-fuel mixture or 41559. lb/hr for a waste input flow rate of 5000. lb/hr.

Input the percent of excess air: ....

If Calculational Mode 2 is being used (i.e., the user inputs the desired operating temperature and *HWI* is to calculate the required amount of excess air), only the stoichiometric air requirement is given; the prompt for the percent of excess air does not appear.

If the air is preheated (for example, by means of a heat exchanger), the resulting incinerator temperature will be higher for the same amount of input air. If the air is much cooler than normal ambient temperatures (around 70°F), this will also affect the incinerator temperature. Air temperatures significantly different from 70°F in either direction (hotter or colder) should be inputted here.

If the air is preheated, input its temperature ^{o}F. (If not, simply depress the RETURN or ENTER key.) ....

---

5. <u>Waste heating value</u>. The net heating value (NHV) of a material is the amount of heat that is released when the material is completely combusted with oxygen. The NHV of the waste may be either inputted by the user or estimated by Dulong's method (cf. Chapter 2 of the *Reference Manual*, Part 4 of this book). Dulong's method is approximate and is recommended only if heat of combustion data are not known. The method is satisfactory only for wastes that contain carbon, hydrogen, oxygen, sulfur, nitrogen and/or chlorine. If this method is used with wastes containing fluorine, bromine and/or iodine, a warning will appear on the screen, but the calculations will continue; in this case, the effect of these three elements on the NHV is ignored. As mentioned earlier, if water is present in the waste and Dulong's method is to be used to estimate the waste's net heating value, the water contribution must be inputted as **WA** during the composition input step. *HWI* assumes that any hydrogen present in the waste (except the hydrogen indirectly inputted as part of the WA) is combustible. This, of course, is not true if some of the hydrogen is already combined with oxygen in the water.

If NHV data are available, and if one of the componential options (Option 1 or 2) was used to input the waste composition, the user has a choice of inputting the NHVs of each component or the NHV of the waste as a whole. If one of the elemental options (Option 3 or 4) was used, only the overall NHV may be inputted.

Net Heating Value:

One of the following methods may be used to supply the required net heating value of the waste-fuel mixture for the calculations. Type:

"1" *if Dulong's method is to be used to calculate the net heating value. (For wastes containing C, H, O, S, N and/or Cl.)*

"2" *if the net heating value of the total waste-fuel mixture is to be inputted.*

"3" *if the net heating values of the individual components of the waste-fuel mixture are to be inputted.*

If Option 1 or 2 was used earlier to input the waste composition, the third choice shown in the above frame would not appear and an input of "3" would not be accepted.

If the net heating value is to be inputted by the user (either "2" or "3" above), the user is given the choice of several different units for the NHV input(s).

The net heating value may be inputted using any one of the following units. Type:

"1" *for Btu/lb*

"2" *for cal/g*

"3" *for Btu/lbmol*

"4" *for cal/gmol*

In the following example, Composition Input Option 4 was used to input the composition of a waste-mixture consisting of 3 components: chlorobenzene, benzene and water. The net heating value of each component is to be inputted individually (This is Choice "3" for the method of inputting the NHV) and units of cal/gmol (Choice "4" for the NHV units) are to be employed.

Note: For purposes of this program, the net heating value is defined as POSITIVE when heat is released during the incineration process and NEGATIVE when heat is absorbed. The overall net heating value of the waste-fuel mixture must be positive, although the net heating value of a particular component may be negative.

Input the NHV of COMPONENT 1 in cal/gmol: 714361

Input the NHV of COMPONENT 2 in cal/gmol: 749426

Input the NHV of COMPONENT 3 in cal/gmol: .... -10519

The "heat of combustion" or NHV for water (inputted as -10519) is actually the heat of vaporization. In the incinerator, the water simply vaporizes. The negative sign reflects the fact that this process absorbs, rather than releases, heat.

If the net heating values of the individual components are being inputted, as in the above example, and one of the components is either AS (ash) or WA (water), an input prompt for the NHV of the ash or water does not appear and *HWI* supplies the NHV (zero in the case of ash). In the following example, Component 2 is water (inputted as WA) and Component 4 is ash.

Input the NHV of COMPONENT 1 in cal/gmol: 714361

The NHV of the water (COMPONENT 2) is -10519 cal/gmol.

Input the NHV of COMPONENT 3 in cal/gmol: 749426

The NHV of the ash (COMPONENT 4) is zero.

---

6. Heat loss. A portion of the heat generated by the combustion process is always lost through the walls of the incinerator. In practice, this effect is often neglected. If the heat loss is to be taken into account, however, *HWI* provides for it as shown in the next prompt.

Input the incinerator heat loss as a percent of the net heating value of the waste-fuel mixture. (If no heat loss is to be assumed, simply depress the RETURN or ENTER key.) ....

---

7. Preliminary design data.

HWI can also be used for some preliminary incinerator design such as sizing, etc. This section can be completely skipped (if desired) when the program is run in the EXPERT mode (cf. Chapter V). In order to perform these design calculations, either the flue gas superficial velocity or the incinerator length-to-diameter ratio (L/D) must be specified in advance. If neither is known, inputting a "1" in response to the next prompt will cause a value of 3.0 to be assumed for the L/D ratio when HWI is being run.

Preliminary Design Data:

For preliminary design purposes, either the superficial throughput velocity of the flue gas or the length-to-diameter ratio of the incinerator must be determined beforehand. (If neither is known, a length-to-diameter ratio of 3.0 may be assumed.) Type:

"1" if a length-to-diameter ratio of 3.0 is to be assumed.

"2" if the flue gas superficial velocity is to be inputted.

"3" if the length-to-diameter ratio is to be inputted.

If "2" is entered, a prompt requesting the superficial velocity appears next.

Input the superficial velocity, ft/sec: ....

If "3" is entered, the following prompt appears.

Input the length-to-diameter ratio of the incinerator: ....

Another parameter which is required for the design calculations is the *heat release rate*. (The *NHV* is a measure of how much heat is released by the waste; the *heat release rate*, which is a function of the type of incinerator and, to a lesser extent, of the nature of the waste, is a measure of how fast that heat will be released.) In actual practice, the heat release rate is often unknown and a typical (maximum) value of 25,000 Btu/hr-cu ft is often assumed. The next prompt requests a value for the heat release rate.

Input the heat release rate of the waste-fuel mixture in Btu/hr-cu ft. (If the RETURN or ENTER key is depressed without entering a number, a heat release rate of 25,000 Btu/hr-cu ft will be assumed.) ....

---

9. Operating temperature

If Calculation Mode 1 (calculation of the operating temperature) is being used, the input phase of *HWI* is now finished. If Calculation Mode 2 (calculation of the percent of excess air) is being employed, the desired operating temperature is next inputted.

Assuming complete combustion, when zero excess air (stoichiometric air) is used, the highest operating temperature that can theoretically be reached in the incinerator results. *HWI* first calculates this temperature and warns the user not to request a temperature above this maximum. In the example below, the maximum temperature (using no excess air) was calculated to be 3870°F.

Operating Temperature:

The maximum operating temperature (referred to as the "theoretical adiabatic flame temperature") that can be achieved under these conditions (i.e., the waste composition, preheated air temperature, percent heat loss that have been inputted) is 3870°F. Do not input a temperature in excess of this maximum.

Input the incinerator operating temperature, ^{o}F: ....

In this case, if the user enters a number greater than 3870, *HWI* will ignore the input and repeat the prompt.

How to Correct Input Errors to *HWI*

If a mistake has been made when inputting information to *HWI* from the keyboard, there are two ways of correcting the input without terminating and restarting the program. Whichever of the two methods is used depends on whether or not the **RETURN** or **ENTER** key was depressed after the incorrect input was typed. One method involves the use of the **BACKSPACE** key and the second involves a special "backtrack" feature of the *HWI* program.

The **BACKSPACE** Key

Whenever any numerical or alphabetical input is typed at the keyboard, it will not be "read" by *HWI* until the **RETURN** or **ENTER** key is depressed. This feature gives the user a chance to correct the input immediately after an error has been made. To correct or change an input that has just been typed but has not yet been "entered" by depressing the **RETURN** or **ENTER** key, simply depress the **BACKSPACE** key one or more times until the input (or the incorrect portion of it) has been erased. Then type the correct input.

The "Backtrack" Feature

If an input error has been made and "entered" by depressing the **RETURN** or **ENTER** key, a special backtrack feature, that allows the user to "go back" to an earlier point in the program, may be used. To use this feature, a "-1", "-2", or "-3" is typed when any NUMERICAL input is called for. A "-1" causes the program to return to the previous input prompt; a "-2" causes the program to return to some earlier prompt; and a "-3" causes the program to return to the opening prompt (i.e., the Main Menu frame). Suppose, for example, the program is in the NOVICE mode, one of the componential input options (Option 3 or 4) has been used, the total net heat value of the waste has been inputted, and the input prompt for the incinerator heat loss now appears on the screen. If a "-1" is entered at this point, the input prompt for the net heating value appears; if a "-2" is entered, the prompt for the NHV units appears; and if a "-3" is entered, the Main Menu frame appears.

Output From *HWI*

The *HWI* output consists mainly of tables of calculated results that appear on the monitor screen. If the user needs a hard copy of any of this output and has a printer or typewriter connected to the terminal, depressing the **SHIFT** and **PRTSC** keys simultaneously will cause whatever is currently appearing on the screen to be printed out. A more complete record can be obtained by depressing the **CTRL** and **PRTSC** simultaneously at any point in the program. Whatever appears on the screen from that point on (until these two keys are again depressed) will be printed out. These actions do not affect the running of *HWI*.

The first output displays the composition of the waste (or waste-fuel mixture). If Composition Input Option 3 or 4 was used, this information is contained in two tables; the first gives the composition by compound and the second, by element. If Composition Input Option 1 or 2 was used, only the elemental composition table is generated. These tables (or this table) appear on the monitor screen right after the input prompt for the waste flow rate, unless their appearance has been suppressed (EXPERT mode only). The componential table includes the mass and mole percent of each component and the elemental table includes the mass and atom percent of each element. If WA (water) was entered as an element (Composition Input Option 1 or 2), its contribution is included with oxygen and hydrogen in the elemental table. If ash is present, its contribution is included in the mass percent columns, but neglected in the atom or mole percent columns.

The remaining three output tables do not appear until after the last input. The first of these tables gives the composition of the flue gas leaving the incinerator (mass percents, mole percents, etc.). The second gives more information on the flue gas including flow rates, temperature, excess air, SO_3 and Cl_2 concentrations in ppm (parts per million by volume), and HCl mass flow rate. (Note: Federal regulations limit HCl emissions from the stack of an incinerator to 4 lb/hr, unless scrubbing equipment is used to remove HCl from the stack gas. In the latter case, the scrubbing efficiency must be better than 99%.) If ash is present, the particulate loading in gr/acf (grains per actual cubic foot) and gr/dscf (grains per dry standard cubic foot) is also given. The latter is corrected to 50% excess air, that is, the tabulated value is the particulate loading in gr/dscf that would be carried by the flue gas if 50% excess had been used. (Note: Federal regulations also require that stack emissions of particulate matter be limited to 0.08 gr/dscf for the stack gas corrected to 50% excess air.)

An example of a typical "Flue Gas Data" table is shown below. These results were generated from Problem 2 listed at the end of Chapter 3. (Also refer to Chapter 5 of the *Reference Manual*, Part 4 of this book.)

FLUE GAS DATA

Mass flow rate, lb/hr	*79238.*
Volumetric flow rate, acfm	*98298.*
Volumetric flow rate, scfm (60°F)	*16931.*
Volumetric flow rate, dscfm (60°F)	*16090.*
Molar flow rate, lbmol/hr	*2675.*
Temperature, °F	*2559.*
Percent excess air	*75.0*
Sulfur trioxide concentration, ppm	*.0*
Chlorine concentration, ppm	*50.9*
HCl mass flow rate, lb/hr	*2058.6*
Particulate loading, gr/acf	*.00*
Particulate loading, gr/dscf -- corrected to 50% excess air......	*.00*

The final output table to appear (unless it has been suppressed using the EXPERT mode) contains preliminary design information and is shown below. Problem 2 was also used to generate these results.

PRELIMINARY INCINERATOR DESIGN DATA

Heat generation rate, Btu/hr	*.5493E+08*
Volume, cu ft	*2197.*
Diameter, ft	*8.9*
Length, ft	*35.5*
L/D ratio	*4.00*
Superficial velocity, ft/sec	*26.5*
Residence time, sec	*1.34*

If the user at this point wishes to run another problem, it is not necessary to terminate and restart the program. The inputting of a "-3" in response to the prompt underneath the last table will cause the Main Menu frame to reappear and the user may start a new problem input.

Error and Warning Messages in *HWI*

Error and warning messages will appear during the running of the *HWI* program for any number of reasons. An error message usually indicates that some action (such as changing an input, for example) must be taken by the user if the calculations are to continue. Most obvious input errors will simply cause *HWI* to repeat the prompt; in this case, since the error should be apparent to the user, no error message is displayed. Some examples of possible error messages that could appear on the screen the running of *HWI* are:

(1) If the data file, *HWISET.DAT*, which contains the information for running *HWI* in the EXPERT mode (cf. Chapter 5 for details), is either missing or defective, one of the following messages will appear:

****** ERROR READING FILE: "HWISET.DAT" ******

****** ERROR IN FILE: "HWISET.DAT" ******

The appropriate action to take in this case is to terminate the *HWI* run (by depressing the C and CTRL keys simultaneously) and to run the program *HWISET* to generate a new *HWISET.DAT* file.

(2) If the user enters either an invalid chemical symbol (i.e., a symbol unrecognizable to *HWI*), enters the same element symbol twice when using Composition Input Option 1 or 2, or enters the same element symbol twice in the same compound when using Composition Input Option 3 or 4, the following error message will appear:

****** ERROR. INVALID CHEMICAL SYMBOL OR ELEMENT ALREADY INPUTTED. ******

3) If the data inputted by the user results in either an operating temperature or a theoretical adiabatic flame temperature (operating temperature using zero excess air) that is unrealistically high or low, the message below, followed by a suggested remedy, will appear on the screen:

****** ERROR DURING CALCULATIONS ******

(4) Occasionally, *HWI* is unable to converge to a solution. Sometimes this is caused by an inputted excess air percent that is too high (Calculation Mode 1) or an inputted net heating value that is too low. If the message: ***run-time error*** appears on the screen, the program has been automatically terminated by the system. The user should restart the program and either try a lower value for the excess air or check the NHV information that was inputted earlier.

The following *warnings* will be given during the execution of *HWI* if the conditions cited in the warnings apply. These warnings do not cause *HWI* to terminate and may be ignored by the user.

(1) If the waste contains more than 5% ash (by mass), the following warning appears after the waste elemental composition table:

******WARNING*** This waste has an ash content greater than 5%. This may yield inaccurate results.***

(2) If the user indicates that Dulong's method for estimating the net heating value of the waste is to be used with a waste containing fluorine, bromine or iodine, the following warning is given:

******WARNING*** Using Dulong's method with a waste containing elements other than C, H, O, S, N and/or Cl may produce inaccurate results.***

(3) If the user chooses Dulong's method to estimate the net heating value of the waste after inputting water as H_2O instead of as "WA", the following warning is issued:

******WARNING*** Apparently, one of the components in the waste is water. Dulong's method should be used ONLY WHEN the contribution of any water present has been inputted as "WA". See the User's Guide.***

(4) If the calculated residence time for the waste is too short, the following warning appears after the last output table:

******WARNING*** A residence time less than 0.75 seconds is unusual.***

(5) If the calculated superficial velocity of the flue gas is abnormally low or high, the following message is given after the last table:

******WARNING*** A gas superficial velocity greater than 50 or lower that 5 ft/sec is unusual.***

CHAPTER FIVE

The Program *HWISET*

What *HWISET* Does

HWISET is a utility program that allows you to exercise some control over the *HWI* program and, to a certain extent, "tailor" *HWI* for a specific use. When *HWI* is run in the NOVICE mode, a complete set of input prompts appears on the monitor, all intermediate results such as the elemental analysis of the waste on both a mass and mole basis is presented, and all of the *HWI* calculations are performed. When *HWI* is run in the EXPERT mode, many of the input prompts, the displaying of the intermediate results and some of the *HWI* calculations can be suppressed, allowing the user to spend less time at the terminal for a single run.

The set of instructions that "tells" *HWI* which prompts, results or calculations to skip is contained in a data file named *HWISET.DAT*. When *HWI* is run, one of the first operations is the reading of these instructions (*HWISET.DAT*) into memory. *HWI* then behaves accordingly. It is the function of the program *HWISET* to generate the instruction data file.

Once the instruction file *HWISET.DAT* has been created, the *HWI* program will respond to that particular set of instructions each time it is run, unless the instructions are changed by running *HWISET* again. *HWI* does not revert back to the NOVICE mode unless instructed to do so through *HWISET.DAT*.

How to Run *HWISET*

To access *HWISET* type the letters **HWISET** at the DOS prompt (>) and depress the **RETURN** or **ENTER** key. The title frame should appear within a few seconds.

The second frame to appear on the screen briefly describes the *HWISET* program and gives instructions for its use. All of the required data for *HWISET* are inputted in response to questions that appear on the screen. In most cases, the input consists of a single digit number.

The run time for *HWISET* is quite short, but if you wish to abort its run before the end of the program is reached, simply depress the **CTRL** and **C** keys simultaneously. The DOS prompt will immediately appear. This has no effect on the previous version of the instruction file *HWISET.DAT*. The existing version of this file is not replaced by a new version unless the *HWISET* program is run to its normal conclusion.

If an incorrect answer is typed, it may be corrected by using the **BACKSPACE** key. If an incorrect answer is typed and cannot be corrected via **BACKSPACE** because the **RETURN** or **ENTER** key has already been depressed, *HWISET* may recognize this input as an error and repeat the input prompt. Such would be the case if a letter is typed when a number is called for, or a "2" is typed when the only choices are "0" or "1". If *HWISET* does not recognize the error and accepts the incorrect input, the run should be aborted (by depressing the **CTRL** and **C** keys simultaneously) and restarted.

What Information Does *HWISET* Ask For?

The input to *HWISET* is organized into 13 questions or question sets, although not all 13 will necessarily appear on screen. If, for example, in answer to the first question regarding the use of the EXPERT or NOVICE mode, you choose the NOVICE mode, no further questions will appear and *HWISET* terminates. The reason for this should be obvious. When *HWI* is run in the NOVICE mode, all prompts, results and calculations are displayed; nothing is suppressed.

The 13 questions are numbered N00 through N12. Each set is listed below as it appears on the screen and then discussed. To distinguish textual material that appears on the screen during the running of *HWISET* from explanatory text, the *HWISET* text is printed in ***bold italics***.

Novice vs. Expert Mode

N00. Do you wish the program set to NOVICE or EXPERT mode?

"0" sets the program to NOVICE mode.

"1" sets the program to EXPERT mode.

Type either "0" or "1": .... N00 =

If a "0" is typed, *HWISET* immediately terminates. If a "1" is typed, the screen clears and Question N01 appears.

Note that, if the answer to any question is zero, it is not necessary to type a "0" and enter it using the **RETURN** or **ENTER** key. By simply depressing the **RETURN** or **ENTER** key without typing anything, a zero is automatically entered. The user will find that this technique will save time, particularly when most of the answers are zero (which is often the case).

N01. Do you wish to preselect which Main Menu Option is to be used? The choices are:

"0" ***No. The Main Menu will appear during program execution.***

"1" ***Yes. Main Menu Option 1 will be used (i.e., incinerator operating temperature will be calculated by HWI). The Main Menu will not appear.***

"2" ***Yes. Main Menu Option 2 will be used (i.e., the percent of excess air required will be calculated by HWI). The Main Menu will not appear.***

Type "0", "1" or "2": .... N01 =

The Main Menu of *HWI* offers a choice between the two calculational modes. Typing a "1" or "2" will cause the suppression of the "Main Menu" frame the next time (or next several times, until *HWISET* is run again) *HWI* is run. In this case, there is no need for the Main Menu since the choice of the calculational mode has already been made.

Waste Composition Input Menu

N02. Do you wish to preset the manner in which the waste composition is to be inputted? The choices are:

"0" ***No. The waste-fuel composition input menu will appear during program execution.***

"1" ***Yes. Prompts for elemental composition, atom basis, will appear; the menu will not.***

"2" Yes. Prompts for elemental composition, mass basis, will appear; the menu will not.

"3" Yes. Prompts for componential composition, mole basis, will appear; the menu will not.

"4" Yes. Prompts for componential composition, mass basis, will appear; the menu will not.

Type "0", "1", "2", "3" or "4": .... N02 =

A "1", "2", "3" or "4" suppresses the appearance of the Waste Composition Input Menu during the next running of *HWI*.

N03. Do you wish to preset the units to be used for the waste-fuel rate? The choices are:

"0" No. The prompt for choosing the units will appear during program execution.

"1" Yes. The units are to be lb/hr. The prompt for choosing units will not appear.

"2" Yes. The units are to be lbmol/hr. The prompt for choosing units will not appear.

Type "0", "1" or "2": .... N03 =

If a "1" or "2" was typed in answer to Question N02, i.e., if the user intends to input the waste composition to *HWI* on an elemental rather than a componential basis, only choices "0" or "1" will appear for Question N03. Note that the *lbmol* unit is meaningless and cannot be used unless the compounds making up the waste are known.

Waste Composition Table

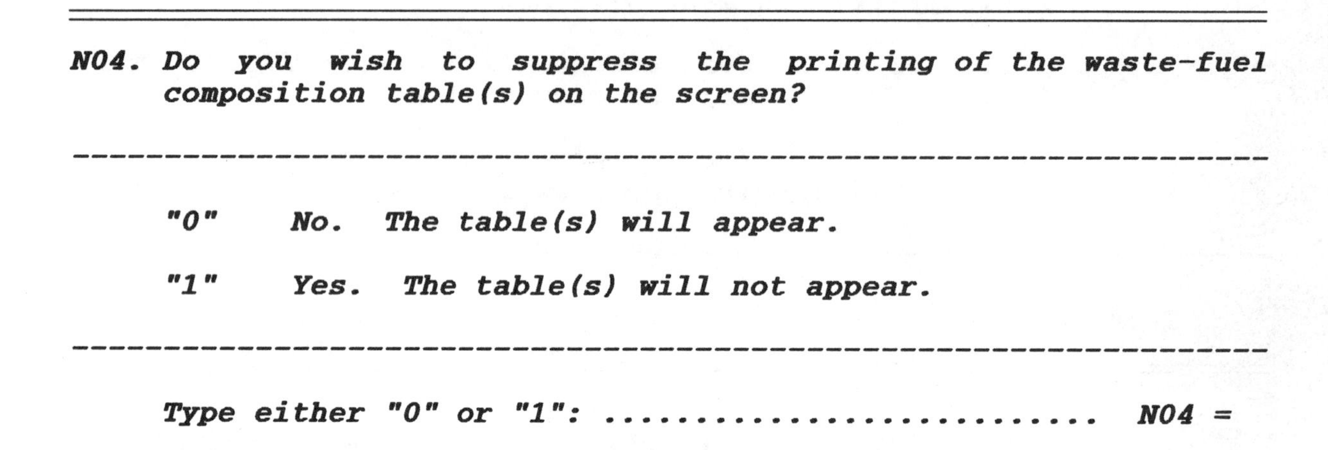

N04. Do you wish to suppress the printing of the waste-fuel composition table(s) on the screen?

"0"	*No. The table(s) will appear.*
"1"	*Yes. The table(s) will not appear.*

Type either "0" or "1": .... N04 =

There are either two tables or one table referred to here, depending on whether a componential or elemental basis is being used for the waste composition input (cf. Chapter 4). If a componential input is chosen, two tables, one showing the waste-fuel composition in terms of the compounds present, the other in terms of the elements present, are presented on the screen during the *HWI* run. If an elemental input is chosen, only the elemental composition table is presented.

Excess Air

If the answer given to Question N01 was "2", i.e., the amount of excess air required to achieve a specified incinerator operating temperature is to be calculated by *HWI*, Question N05 below does not appear on the screen.

N05. Do you wish to preset the input value of the percent excess air? The choices are:

"0"	*No. A prompt for the percent excess air will appear during program execution.*
"1"	*Yes. A prompt will not appear during program execution; the percent excess air is to be preset. (A prompt for inputting this present value will appear next.)*

Type either "0" or "1": .... N05 =

If "1" is chosen, the following prompt appears:

Input the percent excess air: ....

The "%" symbol should not be typed when answering this prompt.

Preheated Air Temperature

N06. Do you wish to suppress the prompt for preheated air?

"0"   No. The prompt will appear during program execution.

"1"   Yes. The prompt will not appear. It will be assumed that the air is not preheated.

Type either "0" or "1": .... N06 =

Net Heating Value (Method)

The number of options allowed for Question N07 depends on the answer given to Question N02. If a "1" or "2" was typed in response to Question N02, i.e., if you intend to input the waste composition to *HWI* on an elemental rather than a componential basis, only choices "0", "1", and "2" will appear for Question N07.

N07. Do you wish to preset the method for inputting or calculating the net heating value (NHV)? The choices are:

"0"   No. A prompt for choosing the method will appear during program execution.

"1" *Yes. Dulong's method will be used to calculate the NHV.*

"2" *Yes. The overall NHV for the waste-fuel mixture will be inputted; a prompt for this input will appear during execution.*

"3" *Yes. The NHV of each component of the waste-fuel mixture will be inputted; prompts for these inputs will appear during program execution.*

Type "0", "1", "2" or "3": .... N07 =

Dulong's approximate method for calculating the NHV should not be used for wastes containing elements other than carbon, hydrogen, oxygen, nitrogen, sulfur and chlorine.

Net Heating Value (Units)

If a "1" was typed in answer to Question N07, i.e., Dulong's method is to be used to calculate the NHV, Question N08 will not appear on the screen.

N08. Do you wish to preset the units to be used for inputting the NHV? The choices are:

"0" *No. A prompt for choosing these units will appear during program execution.*

"1" *Yes. The units will be Btu/lb.*

"2" *Yes. The units will be cal/g.*

"3" *Yes. The units will be Btu/lbmol.*

"4" *Yes. The units will be cal/gmol.*

Type "0", "1", "2", "3" or "4": .... N08 =

If a "1" or "2" was typed in response to Question N02, only choices "0", "1" and "2" will appear for Question N08. Once again, the *lbmol* or *gmol* unit is meaningless unless the waste composition is being inputted on a componential basis.

Heat Loss

N09. Do you wish to preset the value for the percent heat loss (i.e., percent of net heating value) from the incinerator? The choices are:

"0" ***No. A prompt for the percent heat loss will appear during program execution.***

"1" ***Yes. No prompt will appear; it will be assumed that there is no heat loss.***

"2" ***Yes. No prompt will appear and the percent heat loss is to be preset. (A prompt for inputting the percent heat loss will appear next.)***

Type "0", "1" or "2": .... N09 =

If "2" is chosen, the following prompt appears:

Input the heat loss as a percent of the net heating value: ....

Preliminary Design Calculations

N10. Do you wish to omit the Preliminary Design calculations?

"0" ***No. Prompts for these inputs will appear.***

"1" ***Yes. Prompts for these inputs will not appear.***

Type either "0" or "1": .... N10 =

If "1" is chosen, *HWISET* immediately terminates. During subsequent *HWI* runs, none of the design prompts will appear and the table entitled "Preliminary Incinerator Design Data" will not be presented on the screen.

If "0" is chosen, Question N11 appears in the next frame.

L/D Ratio and Superficial Velocity

N11. Do you wish to preset which design parameter is to be specified? The choices are:

"0" ***No. A prompt for choosing this parameter will appear during program execution.***

"1" ***Yes. The length-to-diameter ratio of the incinerator will be assumed to be 3.0.***

"2" ***Yes. The flue gas superficial velocity will be inputted; a prompt for this input will appear during program execution.***

"3" ***Yes. The length-to-diameter ratio of the incinerator will be inputted; a prompt for this input will appear during program execution.***

Type "0", "1", "2" or "3": .... N11 =

Heat Release Rate

As was the case with Question N11, Question N12 appears only if a "0" was typed in response to Question N10.

N12. Do you wish to preset the value of the heat release rate?

"0" *No. A prompt for this input will appear during program execution.*

"1" *Yes. A prompt will not appear; the heat release rate will be assumed to be 25,000 Btu/hr-cu ft.*

"2" *Yes. A prompt will not appear during program execution; the heat release rate is to be preset. (A prompt for inputting this preset value will appear next.)*

Type "0", "1" or "2": .... N12 =

If "2" is chosen, the following and final prompt appears:

Input the heat release rate in Btu/hr-cu ft: ......

After this input, *HWISET* terminates.

How to View the *HWISET.DAT* File

The *HWISET.DAT* file may be viewed on the monitor screen by typing **TYPE HWISET.DAT**, followed by depressing the **RETURN** or **ENTER** key. The file is only five lines long. The first line contains the single-digit answer to Question N00. If this digit is zero, all of the other parameters in the file are also equal to zero. The second line contains the 12 single-digit answers to Questions N01 to N12. Lines 3, 4 and 5 contain the preset values of the percent heat loss, heat release rate and percent excess air, respectively, that were inputted during the last run of *HWISET*. These last three numbers are expressed in scientific or "E" notation. If these variables were not preset during the last *HWISET* run, their values are zero.

It should be pointed out that *HWISET.DAT* can be constructed without the use of the *HWISET* program. Almost any suitable text editor could accomplish this task. However, setting up the *HWISET.DAT* file through the use of *HWISET* is, for most users, probably the more efficient and more convenient method to use.

PART 4

REFERENCE MANUAL FOR THE *HWI SOFTWARE*

The purpose of the *Reference Manual* is to provide the user with a review of the basic principles underlying hazardous waste incineration calculations. The manual is divided into five chapters: 1: An introduction; 2: calculation principles; 3: program details; 4: a chlorobenzene example; and 5: a DDT (dichlorodiphenyltrichloroethane) example. A brief description of the topics covered in each part is given in the Introduction.

CHAPTER ONE

Introduction

The purpose of this reference manual is to provide the user with a review of the basic principles underlying hazardous waste incineration and to present the three key calculations associated with this operation. These include:

(1) thermochemical calculations, which relate operating incinerator temperature, excess air, and feed (including not only waste but also possible fuel requirements) heating value;

(2) stoichiometric calculations, which yield the composition and flow rates of gaseous emissions from incinerators burning (hazardous) waste; and

(3) incinerator design.

The manual is divided into five chapters: (1) this introduction, (2) calculation principles, (3) program details, (4) a chlorobenzene example, and (5) a DDT (dichlorodiphenyl-trichloroethane) example. In *Chapter 2*, calculation principles are reviewed. A broad number of subjects are covered here, including: conservation laws, stoichiometric calculations, and thermodynamics -- including heat capacity, enthalpy, enthalpy of reaction, heating values, thermochemical calculations, chemical reaction equilibrium -- and incinerator design. *Chapter 3* is concerned with program details. The approach employed in the computer calculations is presented, justified, and discussed. The manual concludes with Chapters 4 and 5; the incineration of a chlorobenzene mixture containing sulfur is given in *Chapter 4* and serves as the first tutorial example, and the incineration of a DDT-water mixture is provided in *Chapter 5*, and serves as the second tutorial example. Specific details regarding the use of the programs can be found in the User's Guide (Part 3 of this book) that serves as a companion to this manual.

CHAPTER TWO

Calculation Concepts

In order to completely understand the design as well as the operation and performance of hazardous waste incinerators, it is necessary to first understand the concepts underlying this technology. How can one predict what products will be emitted from effluent streams? At what temperature must the incinerator be operated to ensure compliance with the four nines (99.99% destruction and/or removal efficiency)? How much energy in the form of heat is given off during combustion? Is the waste feed heating value high enough, or must additional fuel be added to assist in the combustion process? If so, how much fuel must be added? The answers to these questions are rooted in the various theories of the conservation laws, thermodynamics, thermochemistry and chemical reaction equilibrium, and will be provided in this and Chapter 3 of this manual. Thus, Chapter 2 discusses a number of scientific principles, primarily from the fields of chemistry, physics, and chemical engineering. Important calculational procedures are presented; the more important among these are methods for predicting (a) incinerator operating temperature based on the properties of the waste or waste-fuel mixture and the amount of excess air employed during combustion, (b) flue products produced when burning a given waste-fuel mixture in stoichiometric or excess air, and (c) the physical design of the incinerator.

CONSERVATION OF MASS

The Conservation Law for Mass can be applied to any process or system. The general form of this law is given by Equation (2-1).

$$\begin{matrix}\text{mass} \\ \text{in}\end{matrix} - \begin{matrix}\text{mass} \\ \text{out}\end{matrix} + \begin{matrix}\text{mass} \\ \text{generated}\end{matrix} = \begin{matrix}\text{mass} \\ \text{accumulated}\end{matrix}$$

or on a time rate basis by

$$\begin{matrix}\text{rate of} \\ \text{mass} \\ \text{in}\end{matrix} - \begin{matrix}\text{rate of} \\ \text{mass} \\ \text{out}\end{matrix} + \begin{matrix}\text{rate of} \\ \text{mass} \\ \text{generated}\end{matrix} = \begin{matrix}\text{rate of} \\ \text{mass} \\ \text{accumulated}\end{matrix} \qquad (2\text{-}1)$$

This equation may be applied either to the total mass involved or to a particular species, on either a mole or mass basis. In incineration processes, it is often necessary to obtain quantitative relationships by writing mass balances on the various elements in the system. This concept is extended in a later section on stoichiometry.

CONSERVATION OF ENERGY[1]

A presentation of the Conservation Law for Energy would be incomplete without a brief review of some introductory thermodynamic principles. Thermodynamics is defined as that science which deals with the relationships among the various forms of energy. A system may possess energy due to its

1. temperature
2. velocity
3. position
4. molecular structure
5. surface
   etc.

The energies corresponding to these states are

1. internal
2. kinetic
3. potential
4. chemical
5. surface
   etc.

Engineering thermodynamics is founded on three basic laws. Energy, like mass and momentum, is conserved. Application of the Conservation Law for Energy gives rise to the First Law. This law, in steady-state equation form for flow processes, is given by

$$\Delta H = Q - W_s \quad (2\text{-}2)$$

where potential, kinetic, etc., energy effects have been neglected and

Q = energy in the form of heat transferred across the boundaries of the system

W_s = energy in the form of mechanical work transferred across the boundaries of the system

ΔH = enthalpy change of the system.

Note: In this manual, properties on either a total (system) or molar basis are represented by upper case letters (e.g., H, E, V, C_P, etc.) while properties on a mass basis are represented by lower case letters (e.g., h, e, v, c_P, etc.).

Perhaps the most important thermodynamic function the engineer works with is the *enthalpy*. The enthalpy is defined by the equation

$$H = E + PV$$

where P = pressure of the system
V = volume of the system.

The terms E and H (and free energy G, to be discussed later) are *state* or *point* functions. By fixing a certain number of variables upon which the function depends, the numerical value of the function is automatically fixed; i.e., it is single-valued. For example, fixing the temperature and pressure of a one-component single-phase system immediately specifies the enthalpy.

The change in enthalpy as it undergoes a change in state from (T_1,P_1) to (T_2,P_2) is given by

$$\Delta H = H_2 - H_1 \tag{2-3}$$

Note that ΔH is independent of the path. This is a characteristic of all state or point functions, i.e., the state of the system is independent of the path by which the state is reached. The terms Q and W_s in Equation (2-2) are path functions; their values depend on the path used between the two states; unless a process or change of state is occurring, path functions have no value.

For a mathematical representation of the above thermodynamic point function, the following can be written:

$$H = H(T,P) \tag{2-4}$$

By the rules of partial differentiation, a differential change in H is given by

$$dH = (\partial H/\partial T)_P dT + (\partial H/\partial P)_T dP \tag{2-5}$$

The term $(\partial H/\partial P)_T$ is assumed to be negligible in most engineering applications. It is exactly zero for an ideal gas and is small for solids and liquids, and gases near ambient conditions. The term $(\partial H/\partial T)_P$ is defined as the heat capacity at constant pressure.

$$C_P = (\partial H/\partial T)_P \tag{2-6}$$

Equation (2-5) may now be written as

$$dH = C_P \, dT \tag{2-7}$$

If average molar heat capacity data are available, the above equation may be integrated to yield

$$H = \overline{C_P} \; T \tag{2-8}$$

where $\overline{C_p}$ = average value of C_p in the temperature range ΔT.

Average molar heat capacity data are provided in Table (2-1) for a number of flue gas (combustion) products. Thus, the calculation of enthalpy change(s) associated with a temperature change may be accomplished through the application of either Equation (2-3) or (2-8). Actual enthalpy data as provided in Table (2-2) must be available for use with Equation (2-3). If the heat capacity is a function of temperature, e.g.,

$$C_p = \alpha + \beta T + \gamma T^2, \qquad (2\text{-}9)$$

Equation (2-7) may be integrated directly between some reference or standard temperature, T_0, and the final temperature, T_1. Then

$$\begin{aligned} H &= H_1 - H_0 \\ &= \alpha(T_1 - T_0) + (\beta/2)(T_1^{\,2} - T_0^{\,2}) + (\gamma/3)(T_1^{\,3} - T_0^{\,3}) \end{aligned} \qquad (2\text{-}10)$$

Equation (2-7) may also be integrated if the heat capacity (as a function of temperature) is of the form

$$C_p = a + b\,T + c\,T^{-2} \qquad (2\text{-}11)$$

The enthalpy change is then given by

$$H = a(T_1 - T_0) + (b/2)(T_1^{\,2} - T_0^{\,2}) - c[(1/T_1) - (1/T_0)] \qquad (2\text{-}12)$$

Tabulated values of α, β, γ and a,b,c for a host of compounds (including some chlorinated organics) are given in Tables (2-3) and (2-4), respectively. The reader is cautioned that Equations (2-9) through (2-12) require the use of the absolute temperature in kelvins.

Most hazardous waste incineration facilities operate in a steady-state flow mode with no significant mechanical or shaft work added (or withdrawn) from the system. For this condition, Equation (2-2) reduces to

$$Q = \Delta H \qquad (2\text{-}13)$$

This equation is routinely used in many hazardous waste incineration calculations. If a unit or system is operated adiabatically, Q = O and Equation (2-13) becomes

$$\Delta H = 0 \qquad (2\text{-}14)$$

Table (2-1). Mean molar heat capacities of gases for the Temperature Range 0°F to T[a]

C_P = Btu/lbmol-°F.

T (°F)	N_2	O_2	H_2O	CO_2	H_2	CO	CH_4	SO_2	NH_3	HCl	NO
0	6.94	6.92	7.93	8.50	6.86	6.92	8.25	9.9	8.80	6.92	7.1
200	6.96	7.03	8.04	9.00	6.89	6.96	8.42	10.0	8.85	6.96	7.2
400	6.98	7.14	8.13	9.52	6.93	7.00	9.33	10.3	9.05	7.01	7.2
600	7.02	7.26	8.25	9.97	6.95	7.05	10.00	10.6	9.40	7.05	7.3
800	7.08	7.39	8.39	10.37	6.97	7.13	10.72	10.9	9.75	7.10	7.3
1000	7.15	7.51	8.54	10.72	6.98	7.21	11.45	11.2	10.06	7.15	7.4
1200	7.23	7.62	8.69	11.02	7.01	7.30	12.13	11.4	10.43	7.19	7.5
1400	7.31	7.71	8.85	11.29	7.03	7.38	12.78	11.7	10.77	7.24	7.6
1600	7.39	7.80	9.01	11.53	7.07	7.47	13.38	11.8		7.29	7.7
1800	7.46	7.88	9.17	11.75	7.10	7.55		12.0		7.33	7.7
2000	7.53	7.96	9.33	11.94	7.15	7.62		12.1		7.38	7.8
2200	7.60	8.02	9.48	12.12	7.20	7.68		12.2		7.43	7.8
2400	7.66	8.08	9.64	12.28	7.24	7.75		12.3		7.47	7.9
2600	7.72	8.14	9.79	12.42	7.28	7.80		12.4		7.52	8.0
2800	7.78	8.19	9.93	12.55	7.33	7.86		12.5		7.57	8.0
3000	7.83	8.24	10.07	12.67	7.38	7.91		12.5		7.61	8.1
3200	7.87	8.29	10.20	12.79	7.43	7.95					
3400	7.92	8.34	10.32	12.89	7.48	8.00					
3600	7.96	8.38	10.44	12.98	7.53	8.04					
3800	8.00	8.42	10.56	13.08	7.57	8.08					
4000	8.04	8.46	10.67	13.16	7.62	8.11					
4200	8.07	8.50	10.78	13.23	7.66	8.14					
4400	8.10	8.54	10.88	13.31	7.70	8.18					
4600	8.13	8.58	10.97	13.38	7.75	8.20					
4800	8.16	8.62	11.08	13.44	7.79	8.23					

[a] From E.T. Williams, and R.C. Johnson, *Stoichiometry for Chemical Engineers*, McGraw-Hill, New York, 1958.

Table (2-2). Molar enthalpies of combustion gases[a] in Btu/lbmol

T (°F)	N_2	Air (MW 28.97)	CO_2	H_2O
32	0	0	0	0
60	194.9	194.6	243.1	224.2
77	312.2	312.7	392.2	360.5
100	473.3	472.7	597.9	545.3
200	1,170	1,170	1,527	1,353
300	1,868	1,870	2,509	2,171
400	2,570	2,576	3,537	3,001
500	3,277	3,289	4,607	3,842
600	3,991	4,010	5,714	4,700
700	4,713	4,740	6,855	5,572
800	5,443	5,479	8,026	6,460
900	6,182	6,227	9,224	7,364
1,000	6,929	6,984	10,447	8,284
1,200	8,452	8,524	12,960	10,176
1,500	10,799	10,895	16,860	13,140
2,000	14,840	14,970	23,630	18,380
2,500	19,020	19,170	30,620	23,950
3,000	23,280	23,460	37,750	29,780

[a] K.A. Kobe, and E.G. Long, "Thermochemistry for the Petroleum Industry," *Pet. Refiner*, **28** (11), 127–132, (1949).

Table (2-3). Molar heat capacites[a,b]

Compound	Formula	α	$\beta \times 10^3$	$\gamma \times 10^6$
	Normal Paraffins (gases)			
Methane	CH_4	3.381	18.044	−4.300
Ethane	C_2H_6	2.247	38.201	−11.049
Propane	C_3H_8	2.410	57.195	−17.533
n-Butane	C_4H_{10}	3.844	73.350	−22.655
n-Pentane	C_5H_{12}	4.895	90.113	−28.039
n-Hexane	C_6H_{14}	6.011	106.746	−33.363
n-Heptane	C_7H_{16}	7.094	123.447	−38.719
n-Octane	C_8H_{18}	8.163	140.217	−44.127
Increment per C atom above C_8	—	1.097	16.667	−5.338
	Normal Monoolefins (gases) (1-alkenes)			
Ethylene	C_2H_4	2.830	28.601	−8.726
Propylene	C_3H_6	3.253	45.116	−13.740
1-Butene	C_4H_8	3.909	62.848	−19.617
1-Pentene	C_5H_{10}	5.347	78.990	−24.733
1-Hexene	C_6H_{12}	6.399	95.752	−30.116
1-Heptene	C_7H_{14}	7.488	112.440	−35.462
1-Octene	C_8H_{16}	8.592	129.076	−40.775
Increment per C atom above C_8	—	1.097	16.667	−5.338
	Miscellaneous Gases			
Acetaldehyde[c]	C_2H_4O	3.364	35.722	−12.236
Acetylene	C_2H_2	7.331	12.622	−3.889
Ammonia	NH_3	6.086	8.812	−1.506
Benzene	C_6H_6	−0.409	77.621	−26.429
1,3-Butadiene	C_4H_6	5.432	53.224	−17.649
Carbon dioxide	CO_2	6.214	10.396	−3.545
Carbon monoxide	CO	6.420	1.665	−0.196
Chlorine	Cl_2	7.576	2.424	−0.965
Cyclohexane	C_6H_{12}	−7.701	125.675	−41.584
Ethyl alcohol	C_2H_6O	6.990	39.741	−11.926
Hydrogen	H_2	6.947	−0.200	0.481
Hydrogen chloride	HCl	6.732	0.433	0.370
Hydrogen sulfide	H_2S	6.662	5.134	−0.854
Methyl alcohol	CH_4O	4.394	24.274	−6.855
Nitric oxide[c]	NO	7.020	−0.370	2.546
Nitrogen	N_2	6.524	1.250	−0.001
Oxygen	O_2	6.148	3.102	−0.923
Sulfur dioxide	SO_2	7.116	9.512	3.511
Sulfur trioxide	SO_3	6.077	23.537	−0.687
Toluene	C_7H_8	0.576	93.493	−31.227
Water	H_2O	7.256	2.298	0.283

Table (2-3). *(Continued)*

Compound	Formula	α	$\beta \times 10^3$	$\gamma \times 10^6$
	Clorinated Organic Liquids			
Methyl chloride	CH_3Cl	8.994	10.280	−21.218
Dichloromethane	CH_2Cl_2	13.672	7.616	−33.272
Chloroform	$CHCl_3$	18.453	4.932	−37.923
Carbon tetrachloride	CCl_4	23.777	2.224	−39.558
Ethyl chloride	C_2H_5Cl	15.267	17.060	−48.798
1,1-Dichloroethane	$C_2H_4Cl_2$	20.014	14.315	−54.917
1,1,2,2-Tetrachloroethane	$C_2H_2Cl_4$	28.397	10.338	−66.739
n-Propyl chloride	C_3H_7Cl	20.703	24.070	−69.351
1,3-Dichloropropane	$C_3H_6Cl_2$	24.840	21.810	−68.214
n-Butyl chloride	C_4H_9Cl	26.258	31.208	−89.482
1-Chloropentane	$C_5H_{11}Cl$	31.812	38.345	−109.612
1-Chloroethylene	C_2H_3Cl	14.058	10.939	−40.586
trans-1,2-Dichloroethylene	$C_2H_6Cl_2$	18.050	8.845	−42.981
Trichloroethylene	C_2HCl_3	22.410	6.394	−46.257
Tetrachloroethylene	C_2Cl_4	26.595	4.106	−46.208
3-Chloro-1-propene	C_3H_5Cl	18.728	18.514	−56.495
Chlorobenzene	C_6H_5Cl	28.558	25.786	−116.150
p-Dichlorobenzene	$C_6H_5Cl_2$	33.265	23.025	−116.848
Hexachlorobenzene	C_6Cl_6	50.759	13.286	−115.620

[a] Selected mainly from values given by H.M. Spencer, J. Justice, and G. Flanagan, *J. Am. Chem. Soc.*, **56**, 2311, 1934; **64**, 2511, 1942; **67**, 1859, 1945; *Ind. Eng. Chem.* **40**, 2152, 1948; also from personal notes: L. Theodore and J. Reynolds, 1986.

[b] Constants for the equation $C_P = \alpha + \beta T + \gamma T^2$, where T is in K and C_P is in Btu/lbmol-°F or cal/gmol-°C.

[c] Applicable range 298–1000 K.

Table (2-4). Molar heat capacities[a,b]

Compound	Formula	Temperature Range (K)	a	$b \times 10^3$	$c \times 10^{-5}$
		Inorganic Gases			
Ammonia	NH_3	298–1800	7.11	6.00	−0.37
Bromine	Br_2	298–3000	8.92	0.12	−0.30
Carbon monoxide	CO	298–2500	6.79	0.98	−0.11
Carbon dioxide	CO_2	298–2500	10.57	2.10	−2.06
Carbon disulfide	CS_2	298–1800	12.45	1.60	−1.80
Chlorine	Cl_2	298–3000	8.85	0.16	−0.68
Hydrogen	H_2	298–3000	6.52	0.78	+0.12
Hydrogen sulfide	H_2S	298–2300	7.81	2.96	−0.46
Hydrogen chloride	HCl	298–2000	6.27	1.24	+0.30
Hydrogen cyanide	HCN	298–2500	9.41	2.70	−1.44
Nitrogen	N_2	298–3000	6.83	0.90	−0.12
Nitrous oxide	N_2O	298–2000	10.92	2.06	−2.04
Nitric oxide	NO	298–2500	7.03	0.92	−0.14
Nitrogen dioxide	NO_2	298–2000	10.07	2.28	−1.67
Nitrogen tetroxide	N_2O_4	298–1000	20.05	9.50	−3.56
Oxygen	O_2	298–3000	7.16	1.00	−0.40
Sulfur dioxide	SO_2	298–2000	11.04	1.88	−1.84
Sulfur trioxide	SO_3	298–1500	13.90	6.10	−3.22
Water	H_2O	298–2750	7.30	2.46	0.00
		Chlorinated Organic Liquids			
Methyl chloride	CH_3Cl	298–1800	3.44	23.52	−0.81
Dichloromethane	CH_2Cl_2	298–1800	5.34	27.00	−1.16
Chloroform	$CHCl_3$	298–1800	8.91	27.18	−1.34
Carbon tetrachloride	CCl_4	298–1800	1.39	24.97	−1.36
Ethyl chloride	C_2H_5Cl	298–1800	2.69	46.81	−1.81
1,1-Dichloroethane	$C_2H_4Cl_2$	298–1800	5.99	47.31	−2.00
1,1,2,2-Tetrachloroethane	$C_2H_2Cl_4$	298–1800	1.15	49.83	−2.38
n-Propyl chloride	C_3H_7Cl	298–1800	2.88	66.18	−2.56
1,3-Dichloropropane	$C_3H_6Cl_2$	298–1800	7.35	63.07	−2.51
n-Butyl chloride	C_4H_9Cl	298–1800	3.22	85.65	−3.32
1-Chloropentane	$C_5H_{11}Cl$	298–1800	3.57	10.51	−4.07
1-Chloroethylene	C_2H_3Cl	298–1800	3.70	35.30	−1.48
trans-1,2-Dichloroethylene	$C_2H_6Cl_2$	298–1800	7.13	34.46	−1.55
Trichloroethylene	C_2HCl_3	298–1800	10.76	33.59	−1.64
Tetrachloroethylene	C_2Cl_4	298–1800	14.99	31.13	−1.63
3-Chloro-1-propene	C_3H_5Cl	298–1800	4.26	52.61	−2.07
Chlorobenzene	C_6H_5Cl	298–1800	−1.12	95.63	−4.24
p-Dichlorobenzene	$C_6H_5Cl_2$	298–1800	3.49	92.98	−4.24
Hexachlorobenzene	C_6Cl_6	298–1800	21.53	81.64	−4.13

[a] Selected from K. Kelley, *U.S. Bur. Mines Bull.* **584** (1960); also from personal notes, L. Theodore and J. Reynolds, 1986.

[b] Constants for the equation $C_P = a + bT + cT^{-2}$, where T is in K and C_P is in Btu/lbmol-°F or cal/gmol-°C.

Although the topics of material and energy balances have been covered separately in this section, it should be emphasized that this segregation does not exist in reality. Incineration is invariably accompanied by heat effects, and one must work with both energy and material balances simultaneously. This is discussed in greater detail in Chapter 3 of this manual.

STOICHIOMETRY

Stoichiometry involves the balancing of an equation for a chemical reaction that provides a quantitative relationship among the reactants and products, and is a direct consequence of the Conservation Law for Mass. In the simplest stoichiometric situation, exact quantities of pure reactants are available, and these quantities react completely to give the desired product(s). For example, complete combustion of pure hydrocarbons yields carbon dioxide and water as the reaction products. Consider the combustion of methane in oxygen:

$$CH_4 + O_2 \longrightarrow CO_2 + H_2O \qquad (2\text{-}15)$$

In order to balance the above equation, two molecules of oxygen are needed. This requires that there be four oxygen atoms on the right side of the equation. This is satisfied by introducing two molecules of water as product. The final balanced reaction becomes

$$CH_4 + 2O_2 \longrightarrow CO_2 + 2H_2O \qquad (2\text{-}16)$$

Thus, two molecules (or moles) of oxygen are required to completely combust one molecule (or mole) of methane to yield one molecule (or mole) of carbon dioxide and two molecules (or moles) of water. Note that the numbers of carbon, oxygen and hydrogen atoms on the right-hand side of this equation are equal to those on the left side. The reader should verify that the total mass (obtained by multiplying the number of each molecule by its molecular weight and summing) on each side of the equation is the same.

The terms used to describe a reaction that does not involve stoichiometric ratios of reactants must be carefully defined in order to avoid confusion. If the reactants are not present in formula or stoichiometric ratio, one reactant is said to be *limiting*; the others are said to be *in excess*. Consider the reaction

$$CO + (1/2)O_2 \longrightarrow CO_2 \qquad (2\text{-}17)$$

If the starting amounts are 1 mole of CO and 3 moles of O_2, CO is the limiting reactant, with O_2 present in excess. There are 2.5 moles of excess O_2, because only 0.5 mole is required to combine with the CO. Thus, there is 500 percent excess oxygen present. The percent excess must be defined in relation to the amount of the reactant necessary to react completely with the limiting reactant. Thus, if for some reason only part of the CO actually reacts, this does not alter the fact that the oxygen is in excess by 500 percent. However, there are often several possible products. For instance, the reactions

$$C + O_2 \text{ -----> } CO_2$$

and

$$C + (1/2)O_2 \text{ -----> } CO \qquad (2\text{-}18)$$

can occur simultaneously. In this case, if there are 3 moles of oxygen present per mole of carbon, the oxygen is in excess. The extent of this excess, however, cannot be definitely fixed. It is customary to choose one product (e.g., the desired one) and specify the excess reactant in terms of this product. For this case, there is 200 percent excess oxygen for the reaction to CO_2, and there is 500 percent excess oxygen for reaction to CO. The discussion on excess oxygen can be extended to excess air using the same approach. Zero percent excess oxygen (or air) is defined as *stoichiometric* or *theoretical* oxygen (or air). This is an important concept since hazardous waste incinerators (and incinerators in general) operate with excess air. For example, approximately 25-50% excess air is employed with liquid injection incinerators. Minimum excess air requirements for rotary kilns are approximately 100% if only bulk solids are burned and 150% if containerized wastes are incinerated.

Consider the combustion of one mole of ethane. The equation for complete combustion may be written

$$C_2H_6 + (7/2)O_2 \text{ -----> } 2CO_2 + 3H_2O \qquad (2\text{-}19)$$

Thus, 2 moles of CO_2 and 3 moles of H_2O will be formed from the complete combustion of 1 mole of C_2H_6. The oxygen required is 3.5 moles. If 60% excess is used, this means an additional 2.1 moles, or a total of 5.6 moles of oxygen is required. Equation (2-19) describes a gas phase reaction. In accordance with Charles' Law, one may also intepret this equation as follows: when one cubic foot of C_2H_6 reacts with 3.5 cubic feet of O_2, 2 cubic feet of CO_2 and 3 cubic feet of H_2O will form. Similar balanced stoichiometric equations may be written for other hydrocarbons and organics to determine oxygen (or air) requirements and the products of combustion. However, this somewhat tedious calculation may sometimes be bypassed by use of Table (2-5).

The above discussion may now be extended to reaction systems that involve the combustion of carbon and/or carbonaceous hazardous wastes and fuels. It will be assumed in the development to follow, for the sake of simplicity, that both

Table (2-5). Combustion constants[a]

Compound	Density	Specific Volume	Heat of Combustion (Btu/ft^3)[c]		Heat of Combustion (Btu/lb)		For 100% Total Air (mole/mole or ft^3/ft^3 of combustible): Required for Combustion			Flue Products			For 100% Total Air (lb/lb of combustible): Required for Combustion			Flue Products			Flammability Limits (% by volume)	
	lb/ft^3[c]	ft^3/lb[c]	Gross (high)	Net (low)	Gross (high)	Net (low)	O_2	N_2	Air	CO_2	H_2O	N_2	O_2	N_2	Air	CO_2	H_2O	N_2	Lower	Upper
Carbon, C[b]	—	—	—	—	14,093	14,093	1.0	3.76	4.76	1.0	—	3.76	2.66	8.86	11.53	3.66	—	8.86	—	—
Hydrogen, H_2	0.0053	187.723	325	275	61,100	51,623	0.5	1.88	2.38	—	1.0	1.88	7.94	26.41	34.34	—	8.94	26.41	4.00	74.20
Oxygen, O_2	0.0846	11.819	—	—	—	—	—	—	—	—	—	—	—	—	—	—	—	—	—	—
Nitrogen (atm), N_2	0.0744	13.443	—	—	—	—	—	—	—	—	—	—	—	—	—	—	—	—	—	—
Carbon monoxide, CO	0.0740	13.506	322	322	4,347	4,347	0.5	1.88	2.38	1.0	—	1.88	0.57	1.90	2.47	1.57	—	1.90	12.50	74.20
Carbon dioxide, CO_2	0.1170	8.548	—	—	—	—	—	—	—	—	—	—	—	—	—	—	—	—	—	—
Paraffin Series																				
Methane, CH_4	0.0424	23.565	1013	913	23,879	21,520	2.0	7.53	9.53	1.0	2.0	7.53	3.99	13.28	17.27	2.74	2.25	13.28	5.00	15.00
Ethane, C_2H_6	0.0803	12.455	1792	1641	22,320	20,432	3.5	13.18	16.68	2.0	3.0	13.18	3.73	12.39	16.12	2.93	1.80	12.39	3.00	12.50
Propane, C_3H_8	0.1196	8.365	2590	2385	21,661	19,944	5.0	18.82	23.82	3.0	4.0	18.82	3.63	12.07	15.70	2.99	1.68	12.07	2.12	9.35
n-Butane, C_4H_{10}	0.1582	6.321	3370	3113	21,308	19,680	6.5	24.47	30.97	4.0	5.0	24.47	3.58	11.91	15.49	3.03	1.55	11.91	1.86	8.41
Isobutane, C_4H_{10}	0.1582	6.321	3363	3105	21,257	19,629	6.5	24.47	30.97	4.0	5.0	24.47	3.58	11.91	15.49	3.03	1.55	11.91	1.80	8.44
n-Pentane, C_5H_{12}	0.1904	5.252	4016	3709	21,091	19,517	8.0	30.11	38.11	5.0	6.0	30.11	3.55	11.81	15.35	3.05	1.50	11.81	—	—
Isopentane, C_5H_{12}	0.1904	5.252	4008	3716	21,052	19,478	8.0	30.11	38.11	5.0	6.0	30.11	3.55	11.81	15.35	3.05	1.50	11.81	—	—
Neopentane, C_5H_{12}	0.1904	5.252	3993	3693	20,970	19,396	8.0	30.11	38.11	5.0	6.0	30.11	3.55	11.81	15.35	3.05	1.50	11.81	—	—
n-Hexane, C_6H_{14}	0.2274	4.398	4762	4412	20,940	19,403	9.5	35.76	45.26	6.0	7.0	35.76	3.53	11.74	15.27	3.06	1.46	11.74	1.18	7.40

Table (2-5). *(Continued)*

Compound	Density	Specific Volume	Heat of Combustion (Btu/ft³)[c]		Heat of Combustion (Btu/lb)		For 100% Total Air (mole/mole or ft³/ft³ of combustible) Required for Combustion			Flue Products			For 100% Total Air (lb/lb of combustible) Required for Combustion			Flue Products			Flammability Limits (% by volume)	
	lb/ft³[c]	ft³/lb[c]	Gross (high)	Net (low)	Gross (high)	Net (low)	O_2	N_2	Air	CO_2	H_2O	N_2	O_2	N_2	Air	CO_2	H_2O	N_2	Lower	Upper
Olefin Series																				
Ethylene, C_2H_4	0.0746	13.412	1614	1513	21,644	20,295	3.0	11.29	14.29	2.0	2.0	11.29	3.42	11.39	14.81	3.14	1.29	11.39	2.75	28.60
Propylene, C_3H_6	0.1110	9.007	2336	2186	21,041	19,691	4.5	16.94	21.44	3.0	3.0	16.94	3.42	11.39	14.81	3.14	1.29	11.39	2.00	11.10
1-Butene, C_4H_8	0.1480	6.756	3084	2885	20,840	19,496	6.0	22.59	28.59	4.0	4.0	22.59	3.42	11.39	14.81	3.14	1.29	11.39	1.75	9.70
Isobutene, C_4H_8	0.1480	6.756	3068	2869	20,730	19,382	6.0	22.59	28.59	4.0	4.0	22.59	3.42	11.39	14.81	3.14	1.29	11.39	—	—
1-Pentene, C_5H_{10}	0.1852	5.400	3836	3586	20,712	19,363	7.5	28.23	35.73	5.0	5.0	28.23	3.42	11.39	14.81	3.14	1.29	11.39	—	—
Aromatic Series																				
Benzene, C_6H_6	0.2060	4.852	3751	3601	18,210	17,480	7.5	28.23	35.73	6.0	3.0	28.23	3.07	10.22	13.50	3.38	0.69	10.22	1.40	7.10
Toluene, C_7H_8	0.2431	4.113	4484	4284	18,440	17,620	9.0	33.88	42.88	7.0	4.0	33.88	3.13	10.40	13.53	3.34	0.78	10.40	1.27	6.75
Xylene, C_8H_{10}	0.2803	3.567	5230	4980	18,650	17,760	10.5	39.52	50.02	8.0	5.0	39.52	3.17	10.53	13.70	3.32	0.85	10.53	1.00	6.00
Miscellaneous Gases																				
Acetylene, C_2H_2	0.0697	14.344	1499	1448	21,500	20,776	2.5	9.41	11.91	2.0	1.0	9.41	3.07	10.22	13.30	3.38	0.69	10.22	—	—
Napthalene, $C_{10}H_8$	0.3384	2.955	5854	5654	17,298	16,708	12.0	45.17	57.17	10.0	4.0	45.17	3.00	9.97	12.96	3.43	0.56	9.97	—	—
Methyl alcohol, CH_3OH	0.0846	11.820	868	768	10,259	9,078	1.5	5.65	7.15	1.0	2.0	5.65	1.50	4.98	6.48	1.37	1.13	4.98	6.72	36.50
Ethyl alcohol, C_2H_5OH	0.1216	8.221	1600	1451	13,161	11,929	3.0	11.29	14.29	2.0	3.0	11.29	2.08	6.93	9.02	1.92	1.17	6.93	3.28	18.95
Ammonia, NH_3	0.0456	21.914	441	365	9,668	8,001	0.75	2.82	3.57	—	1.5	3.32	1.41	4.69	6.10	—	1.59	5.51	15.50	27.00
										SO_2						SO_2				
Sulfur, S[b]	—	—	—	—	3,983	3,983	1.0	3.76	4.76	1.0	—	3.76	1.00	3.29	4.29	2.00	—	3.29	—	—
Hydrogen sulfide, H_2S	0.0911	10.979	647	596	7,100	6,545	1.5	5.65	7.15	1.0	1.0	5.65	1.41	4.69	6.10	1.88	0.53	4.69	4.30	45.50
Sulfur dioxide, SO_2	0.1733	5.770	—	—	—	—	—	—	—	—	—	—	—	—	—	—	—	—	—	—
Water vapor, H_2O	0.0476	21.017	—	—	—	—	—	—	—	—	—	—	—	—	—	—	—	—	—	—
Air	0.0766	13.063	—	—	—	—	—	—	—	—	—	—	—	—	—	—	—	—	—	—
Gasoline	—	—	—	—	—	—	—	—	—	—	—	—	—	—	—	—	—	—	1.40	7.60

[a] Adapted from "Fuel Flue Gases." *Combustion Flame and Explosions of Gases*, American Gas Association, New York, NY, 1951.
[b] Carbon and sulfur are considered as gases for molal calculations only.
[c] All gas volumes corrected to 60°F and 30 in. Hg dry.

the air and mixture are dry. Throughout this manual, air is assumed to contain 21% oxygen by volume. This is perhaps a bit high by a few hundredths of a percent. The remaining 79% is assumed to be inert and to consist of nitrogen (and a trace of the noble gases). The usual incineration products of hydrocarbons are carbon dioxide and water. The combustion products of other organics may contain additional compounds; for example, an organic chloride will also produce hydrochloric acid and/or chlorine. Combustion products from certain fuels will yield sulfur dioxide and nitrogen. So-called complete combustion of an organic or a fuel involves conversion of all the elemental carbon to carbon dioxide, hydrogen to water (unless chlorine is present), sulfur to sulfur dioxide, nitrogen to its elemental form of N_2, etc. Thus, the *theoretical* or *stoichiometric* oxygen described above for a combustion reaction is the amount of oxygen required to burn all the carbon to carbon dioxide, all the hydrogen to water, etc; *excess* oxygen (or *excess* air) is the oxygen furnished over and above the theoretical oxygen required for combustion. Since combustion calculations assume dry air to contain 21% oxygen and 79% nitrogen on a mole or volume basis, 4.76 moles of air consists of 1.0 mole of oxygen and 3.76 moles of nitrogen. The temperature of the products that result from complete combustion under adiabatic conditions is defined as the *adiabatic flame temperature*; this will be discussed in more detail in the next Section.

The balanced reaction equation for the combustion of 1.0 lbmole of chlorobenzene, C_6H_5Cl, in stoichiometric oxygen, is

$$C_6H_5Cl + 7\ O_2 \longrightarrow 6\ CO_2 + 2\ H_2O + HCl \qquad (2\text{-}20)$$

(The incineration of this hazardous waste is treated in extensive detail in Chapter 4.) Since air, not oxygen, is employed in incineration processes, the above equation with air becomes

$$C_6H_5Cl + 7\ O_2 + 26.3\ N_2 \longrightarrow 6\ CO_2 + 2\ H_2O + HCl + 26.3\ N_2 \qquad (2\text{-}21)$$

where the nitrogen in the air has been retained on both sides of the equation since it does not participate in the combustion reaction. The moles and masses involved in this reaction, based on the stoichiometric combustion of 1.0 lbmole of chlorobenzene, are given below.

	C_6H_5Cl	+ 7 O_2	+ 26.3 N_2	----> 6 CO_2	+ 2 H_2O	+ HCl	+ 26.3 N_2
moles	1	7	26.3	6	2	1	26.3
MW	112.5	32	28	44	18	36.5	28
mass	112.5	224	736.4	264	36	36.5	736.4

initial mass = 1072.9
final mass = 1072.9
initial number of moles = 34.3
final number of moles = 35.3

Note that, in accordance with the Conservation Law for Mass, the initial and final masses balance. The number of moles, as is typically the case in incinerator calculations, do not balance. The concentrations of the various species may also be calculated. For example,

% CO_2 by weight = (264/1072.9) 100% = 24.61%
% CO_2 by mole (or volume) = (6/35.3) 100% = 17.0%
% CO_2 by weight (dry basis) = (264/1036.9) 100% = 25.46%
% CO_2 by mole (dry basis) = (6/33.3) 100% = 18.0%

The air requirement for the above reaction is 33.3 lbmoles. This is stoichiometric or 0% excess air (EA). For 100% EA (100% above stoichiometric), one would use (33.3)(2.0) or 66.6 lbmoles of air. For 50% EA, one would use (33.3)(1.5) or 50 lbmoles of air; for this condition 16.7 lbmoles excess or additional air is employed.

If 100% excess air is employed in the incineration of 1.0 lbmole of chlorobenzene, the combustion equation would become

$$C_6H_5Cl + 14\ O_2 + 52.6\ N_2$$
$$\text{----->}\ 6\ CO_2 + 2\ H_2O + HCl + 7\ O_2 + 52.6\ N_2 \qquad (2\text{-}22)$$

The reader is left with the exercise of verifying the results below:

total final number of moles = 68.6
%O_2 by mole (or volume) = (7/68.6) 100% = 10.2%
%HCl by mole = (1/68.6) 100% = 1.46%

THERMOCHEMISTRY

Consider now the energy effects associated with a chemical reaction. To introduce this subject, the reader is reminded that the engineer and applied scientist are rarely concerned with the magnitude or amount of the energy in a system; the primary concern is with changes in the amount of energy. In measuring energy changes for systems, the enthalpy, H, has been found to be the most convenient term to work with. There are many different types of enthalpy effects; these include:

1. sensible (temperature)
2. latent (phase)
3. reaction (chemical)
4. dilution (with water), e.g., HCl with H_2O

5. solution (nonaqueous), e.g., HCl with a solvent other than water.

The sensible enthalpy change (1) was reviewed in an earlier section. The latent enthalpy change (2) finds application in hazardous waste incineration calculations in determining the heat (enthalpy) of condensation or vaporization of water. Steam tables (or the equivalent) are usually employed for this determination. The dilution (4) and solution (5) enthalpy effects are often significant in some industrial absorber calculations, but can safely be neglected in hazardous waste incineration calculations. The *heat of reaction* (3) is defined as the enthalpy change of a system undergoing chemical reaction. If the reactants and products are at the same temperature and in their standard states, the heat of reaction is termed the *standard heat of reaction*. For engineering purposes, the standard state of a chemical may be taken as the pure chemical at 1 atm pressure. A superscript zero is often employed to identify a standard heat of reaction, e.g., ΔH^o. A T subscript (ΔH^o_T) is sometimes used to indicate the temperature; standard heat of reaction data are meaningless unless the temperatures are specified. ΔH^o_{298} data (i.e., for 298 K or 25°C) for many reactions are available in the literature.[2,3] These enthalpy effects play a significant role in many heat effect calculations. As described earlier, the First Law of Thermodynamics provides that, in a steady flow process with no mechanical work,

$$Q = \Delta H \qquad (2\text{-}23)$$

The heat of formation, ΔH_f, is defined as the enthalpy change occurring during a chemical reaction where one mole of a product is formed from its elements. The standard enthalpy of formation, ΔH_f^o, is applied to formation reactions that occur at constant temperature with each element and the product in its standard state.

Consider the formation equation for CO_2 at standard conditions at 25°C.

$$C + O_2 \xrightarrow{\Delta H^o_{f\,298}} CO_2 \qquad (2\text{-}24)$$

Enthalpy: $\quad 0 \quad 0 \qquad\qquad H_{CO2}$

Once again, this equation reads: "1 mole of carbon and 1 mole of oxygen react to form 1 mole of carbon dioxide." The enthalpies of reactants and products are printed below the symbols in the equation, and the enthalpy change for the reaction, ΔH_f^o, placed above the arrow. Note that the enthalpies of elements in their standard states (pure, 1 atm) at 25°C have arbitrarily been set equal to zero. The enthalpy change accompanying this reaction is the *standard heat of formation* and is given by

$$\Delta H_{f\ 298}^{o} = H^{o}_{CO_2} - (H^{o}_{C} + H^{o}_{O_2}) = H^{o}_{CO_2} \qquad (2\text{-}25)$$

The *standard heat of combustion* at temperature T is defined as the enthalpy change during a chemical reaction in which one mole of material is burned in oxygen, where all reactants and products are in their standard states. This quantity finds extensive application in calculating enthalpy changes for incineration reactions, and is often given in the literature for 60°F (16°C). Although much of the literature data on standard heats of reaction are given for 25°C (76°F), there is little sensible enthalpy difference between these two temperatures and the two sets of data may be considered interchangeable.

Note that the above formation reaction for CO_2 is a combustion reaction; the heat of formation, in this case, is therefore equal to the heat of combustion. Since a combustion reaction is one type of chemical reaction, the development to follow will concentrate on chemical reactions in general.

Chemical (stoichiometric) equations may be combined by addition or subtraction; the standard heat of reaction, ΔH^{o}, associated with each equation may likewise be combined to give the standard heat of reaction associated with the resulting chemical equation. This is possible, once again, because enthalpy is a point function, and these changes are independent of path. In particular, formation equations and standard heats of formation may always be combined by addition and subtraction to produce any desired equation and its accompanying standard heat of reaction. This desired equation cannot itself be a formation equation. Thus, the enthalpy change for a chemical reaction is the same whether it takes place in one or several steps. This is referred to as the *Law of Constant Enthalpy Summation* and is a direct consequence of the First Law.

Consider the general reaction equation

$$aA + bB + \ldots \longrightarrow cC + dD + \ldots \qquad (2\text{-}26)$$

where A,B = formulas for the reactants, R

C,D = formulas for the products, P

a,b,c,d = the stoichiometric coefficients of the balanced equation.

To simplify the presentation that follows, Equation (2-26) is shortened to

$$aA + bB \longrightarrow cC + dD \qquad (2\text{-}27)$$

(Although this presentation plus those in later sections will deal with the hypothetical species, A,B,C, etc., application to real systems can be found in the two illustrative examples presented in Chapters 4 and 5 of Part 4). This reaction equation reads: "a moles of A react with b moles of B to form c moles of C and d moles of D." The heat of reaction for this chemical change is given by

$$\Delta H^o = cH^o_C + dH^o_D - aH^o_A - bH^o_B \quad (2\text{-}28)$$

where H^o_i = enthalpy of component *i* in its standard state.

If the temperature is 25°C, the enthalpies of the elements in their standard states are, by convention, equal to zero. Therefore

$$H^o_{i,298} = (\Delta H_f^o)_{i,298} \quad (2\text{-}29)$$

Substituting Equation (2-29) for each component in Equation (2-28) yields

$$\Delta H^o_{298} = c(\Delta H_f^o)_C + d(\Delta H_f^o)_D - a(\Delta H_f^o)_A - b(\Delta H_f^o)_B$$

or

$$\Delta H^o_{298} = \sum_P n_P (\Delta H_f^o)_P - \sum_R n_R (\Delta H_f^o)_R \quad (2\text{-}30)$$

where P = products
R = reactants
n_P, n_R = coefficients from the chemical equation

Thus, the standard heat of a reaction is obtained by taking the difference between the standard heat of formation of the products and that of the reactants. If the standard heat of reaction or formation is negative (exothermic), as is the case with most incineration reactions, then energy is liberated as a result of the reaction. Energy is absorbed if ΔH is positive (endothermic). Standard heat of formation and standard enthalpy of combustion data at 25°C are provided in Table (2-6). Both of these heat (or enthalpy) effects find extensive application in incinerator calculations.

Other tables of heat of formation, combustion, and reaction are available in the literature (particularly thermodynamics text/reference books) for a wide variety of compounds.[1,2] It is important to note that these tables are meaningless unless the stoichiometric equations, temperature and the state of the reactants and products are included. However, "*heat of reaction*" is a term rarely employed in air pollution and/or incinerator calculations. The two terms most often used in this field are the *gross* (or *higher*) *heating value* and the *net* (or *lower*) *heating value*. The former is designated by *HHV* or HV_G

Table (2-6). Standard heats of formation and combus-bustion at 25°C in cal/gmol[a,b]

Compound	Formula	State	ΔH°_{f298}	$-\Delta H^\circ_{c298}$
	Normal Paraffins			
Methane	CH_4	g	−17,889	212,800
Ethane	C_2H_6	g	−20,236	372,820
Propane	C_2H_8	g	−24,820	530,600
n-Butane	C_4H_{10}	g	−30,150	687,640
n-Pentane	C_5H_{12}	g	−35,000	845,160
n-Hexane	C_6H_{14}	g	−39,960	1,002,570
Increment per C atom above C_6	—	g	−4,925	157,440
	Normal Monoolefins (1-alkenes)			
Ethylene	C_2H_4	g	12,496	337,230
Propylene	C_3H_6	g	4,879	491,990
1-Butene	C_4H_8	g	−30	649,450
1-Pentene	C_5H_{10}	g	−5,000	806,850
1-Hexene	C_6H_{12}	g	−9,960	964,260
Increment per C atom above C_6	—	g	−4,925	157,440
	Miscellaneous Organic Compounds			
Acetaldehyde	C_2H_4O	g	−39,760	
Acetic acid	$C_2H_4O_2$	l	−116,400	
Acetylene	C_2H_2	g	54,194	310,620
Benzene	C_6H_6	g	19,820	789,080
Benzene	C_6H_6	l	11,720	780,980
1,3-Butadiene	C_4H_6	g	26,330	607,490
Cyclohexane	C_6H_{12}	g	−29,430	944,790
Cyclohexane	C_6H_{12}	l	−37,340	936,880
Ethanol	C_2H_6O	g	−56,240	
Ethanol	C_2H_6O	l	−66,356	
Ethylbenzene	C_8H_{10}	g	7,120	1,101,120
Ethylene glycol	$C_2H_6O_2$	l	108,580	
Ethylene oxide	C_2H_4O	g	−12,190	
Methanol	CH_4O	g	−48,100	
Methanol	CH_4O	l	−57,036	
Methylcyclohexane	C_7H_{14}	g	−36,990	1,099,590
Methylcyclohexane	C_7H_{14}	l	−45,450	1,091,130
Styrene	C_8H_8	g	35,220	1,060,900
Toluene	C_7H_8	g	11,950	943,580
Toluene	C_7H_8	l	2,870	934,500
	Miscellaneous Inorganic Compounds			
Ammonia	NH_3	g	−11,040	
Calcium carbide	CaC_2	s	−15,000	
Calcium carbonate	$CaCO_3$	s	−288,450	
Calcium chloride	$CaCl_2$	s	−190,000	

Table (2-6). *(Continued)*

Compound	Formula	State	ΔH°_{f298}	$-\Delta H^\circ_{c298}$
Calcium chloride	$CaCl_2 \cdot 6H_2O$	s	−623,150	
Calcium hydroxide	$Ca(OH)_2$	s	−235,800	
Calcium oxide	CaO	s	−151,900	
Carbon	C	Graphite	—	94,052
Carbon dioxide	CO_2	g	−94,052	
Carbon monoxide	CO	g	−26,416	67,636
Hydrochloric acid	HCl	g	−22,063	
Hydrogen	H_2	g	—	68,317
Hydrogen sulfide	H_2S	g	−4,815	
Iron oxide	FeO	s	−64,300	
Iron oxide	Fe_3O_4	s	−267,000	
Iron oxide	Fe_2O_3	s	−196,500	
Iron sulfide	FeS_2	s	−42,520	
Lithium chloride	$LiCl$	s	−97,700	
Lithium chloride	$LiCl \cdot H_2O$	s	−170,310	
Lithium chloride	$LiCl \cdot 2H_2O$	s	−242,100	
Lithium chloride	$LiCl \cdot 3H_2O$	s	−313,500	
Nitric acid	HNO_3	l	−41,404	
Nitrogen oxides	NO	g	21,600	
	NO_2	g	8,041	
	N_2O	g	19,490	
	N_2O_4	g	2,309	
Sodium carbonate	Na_2CO_3	s	−270,300	
Sodium carbonate	$Na_2CO_3 \cdot 10H_2O$	s	−975,600	
Sodium chloride	$NaCl$	s	−98,232	
Sodium hydroxide	$NaOH$	s	101,990	
Sulfur dioxide	SO_2	g	−70,960	
Sulfur trioxide	SO_3	g	−94,450	
Sulfur trioxide	SO_3	l	−104,800	
Sulfuric acid	H_2SO_4	l	−193,910	
Water	H_2O	g	−57,798	
Water	H_2O	l	−68,317	
	Chlorinated Organic Compounds			
Methyl chloride	CH_3Cl	l	−20,630	
Dichloromethane	CH_2Cl_2	l	−22,800	
Chloroform	$CHCl_3$	l	−24,200	36,900
Carbon tetrachloride	CCl_4	l	−24,000	
Ethyl chloride	C_2H_5Cl	l	−26,700	
1,1-Dichloroethane	$C_2H_4Cl_2$	l	−31,050	
1,1,2,2-Tetrachloroethane	$C_2H_2Cl_4$	l	−36,500	
n-Propyl chloride	C_3H_7Cl	l	−31,100	
1,3-Dichloropropane	$C_3H_6Cl_2$	l	−38,600	
n-Butyl chloride	C_4H_9Cl	l	−35,200	
1-Chloropentane	$C_5H_{11}Cl$	l	−41,800	
1-Chloroethylene	C_2H_3Cl	l	−8,400	
trans-1,2-Dichloroethylene	$C_2H_6Cl_2$	l	−1,000	
Trichloroethylene	C_2HCl_3	l	−1,400	

Table (2-6). *(Continued)*

Compound	Formula	State	ΔH°_{f298}	$-\Delta H^\circ_{c298}$
Tetrachloroethylene	C_2Cl_4	l	−3,400	
3-Chloro-1-propene	C_3H_5Cl	l	−150	
Chlorobenzene	C_6H_5Cl	l	12,390	
p-Dichlorobenzene	$C_6H_5Cl_2$	l	5,500	
Hexachlorobenzene	C_6Cl_6	l	−8,100	510,000
Benzylchloride	C_7H_7Cl	l		782,000
1,1,1-trichloro-2,2-bis (*p*-chlorophenyl)ethane (DDT)	$C_{14}H_9Cl_5$	l		1,600,000

[a] Selected mainly from F.D. Rossini, ed., "Selected Values of Physical and Thermodynamic Properties of Hydrocarbons and Related Compounds," *American Petroleum Institute Research Project 44*, Carnegie Institute of Technology, Pittsburgh, PA, 1953; F.D. Rossini, D.D. Wagman, W.H. Evans, S. Levine, and I. Jaffe, "Selected Values of Chemical Thermodynamic Properties," *Natl. Bur. Stand. Circ. 500*, 1952; also from personal notes, L. Theodore and J. Reynolds, 1985.

[b] For combustion reactions the products are H_2O (l) and CO_2 (g).

and the latter by *NHV* or HV_N. The gross heating value represents the enthalpy change or heat released when a compound is stoichiometrically combusted (reacted) at a reference temperature with the final (flue) products also at the same reference temperature and any water present in the liquid state. Most of these data are available at a reference temperature of $60^{o}F$. The net heating value is similar to the gross heating value except the water is in the vapor state. The difference (if any) between the two values represents the energy necessary to vaporize any water present. Thus, the standard heat of reaction and the gross and/or net heating values employed in the incinerator industry both represent the same phenomenon. Gross and net heating values for a number of hydrocarbons are presented in Table (2-5). In addition, the net heating value may be approximated by a form of Dulong's equation that includes the chlorine content:

$$NHV = 14,000\ m_C + 45,000\ (m_H - 0.125\ m_O) - 760\ m_{Cl} + 4500\ m_S \qquad (2\text{-}31)$$

where NHV = net heating value of waste-fuel mixture, Btu/lb

m_i = mass fraction of component *i*

i = subscript representing each element in the waste-fuel mixture

Another term commonly employed in incinerator calculations is the available heat, usually designated as HA_T. The available heat at any temperature T is the gross heating value minus the amount of heat, $\Sigma\Delta H$, required to take the product(s) of combustion (flue gas) from the reference temperature to that temperature T. Thus,

$$HA_T = HHV - \Sigma\Delta H \qquad (2\text{-}32)$$

If all the heat liberated by the reaction goes into heating up the products of combustion (the flue gas), the temperature achieved is defined as the *flame temperature*. If the combustion process is conducted adiabatically, i.e., with no heat transfer to the surroundings, the final temperature achieved by the flue gas is defined as the *adiabatic flame temperature*. If the combustion process is conducted with theoretical or stoichiometric air (zero percent excess), the resulting temperature is defined as the *theoretical adiabatic flame temperature*.

Effect of Temperature on Heats of Reaction

The heat of reaction is a function of temperature because the heat capacities of both the reactants and products vary with temperature. One can describe this effect mathematically in the following manner:

$$\Delta H_T^{o} = \Delta H^{o}_{298} + \int_{298}^{T} \Delta c_p \, dT \qquad (2\text{-}33)$$

where T = absolute temperature, K

$$\Delta c_p = \sum_P nc_p - \sum_R nc_p$$

The heat capacity for each product and reactant can be expressed by either

$$c_p = a + bT + c(T)^{-2}$$

or

$$c_p = \alpha + \beta T + \gamma T^2 \qquad (2\text{-}34)$$

For the former case,

$$\Delta c_p = \Delta a + \Delta b(T) + \Delta c(T)^{-2} \qquad (2\text{-}35)$$

where $$\Delta a = \sum_P na - \sum_R na \qquad (2\text{-}36)$$

with similar definitions for Δb and Δc. The following expression for the standard heat of reaction at temperature T is obtained by combining Equations (2-33) and (2-35):

$$\Delta H^{o}_T = \Delta H^{o}_{298} + \int_{298}^{T} [\Delta a + (\Delta b)T + (\Delta c)T^{-2}]dT$$

or

$$\Delta H^{o}_T = \Delta H^{o}_{298} + \Delta a(T - 298) + (\Delta b/2)(T^2 - 298^2) - \Delta c[(1/T)-(1/298)] \qquad (2\text{-}37)$$

If all the constant terms in this equation are collected and lumped together into a constant designated ΔH_o, the result is

$$\Delta H^{o}_T = \Delta H_o + \Delta aT + (\Delta b/2)T^2 - \Delta c(1/T) \qquad (2\text{-}38)$$

where ΔH^{o}_T = standard heat of reaction at temperature T

$$\Delta H_o = \text{constant} = \Delta H^o_{298} - 298\Delta a - (\Delta b/2)(298)^2 + \Delta c/298$$

Use of the equations in this subsection require that the temperature T be expressed in kelvins.

INCINERATOR TEMPERATURE CALCULATIONS

In order to calculate the operating temperature, fuel requirements, and (excess) air requirements for an incineration operation, one must apply the Conservation Laws for Mass and Energy in conjunction with thermochemical principles. This is an extremely involved, rigorous calculation. An enthalpy balance is applied around the incinerator following a comprehensive overall and componential material balances. The enthalpy balance must account for all temperature changes (effects) of both the feed and fuel (reactants) as well as the flue gas (products). Latent (phase) and combustion (reaction) enthalpy effects must also be included in the analysis.

A rigorous flame temperature calculation is schematically depicted in Figure (2-1). An overall enthalpy approach is now applied to this incinerator calculation. It should be kept in mind that any convenient calculation path can be selected in determining the enthalpy change between reactants (initial state) and products (final state). Since heat (enthalpy) of reaction and/or net (or gross) heating values are available at a reference temperature (T_0) only, the calculations proceed along a chosen path that includes the reaction at this reference temperature. Referring to Figure (2-1), the enthalpy change (gain) associated with the cooling of the reactants (feed) is first obtained. This is followed by the enthalpy changes associated with the combustion reaction, the heating of the flue products and finally, the heat loss from the incinerator. These thermochemical calculations have been developed on a mass basis as is customary in incineration applications. Heating values -- both higher (gross) and lower (net) -- are expressed in Btu/lb.

<u>cooling step</u>:

$$\Delta h_{St} = m_{St}\, c_{PSt}\, (T_0 - T_{St})$$

$$\Delta h_E = m_E\, c_{PE}\, (T_0 - T_E)$$

$$\Delta h_W = m_W\, c_{PW}\, (T_0 - T_W)$$

$$\Delta h_F = m_F\, c_{PF}\, (T_0 - T_F) \qquad (2\text{-}39)$$

where Δh_i = sensible enthalpy change for cooling *i* from initial temperature T_i to reference temperature T_0, Btu/lb mixture

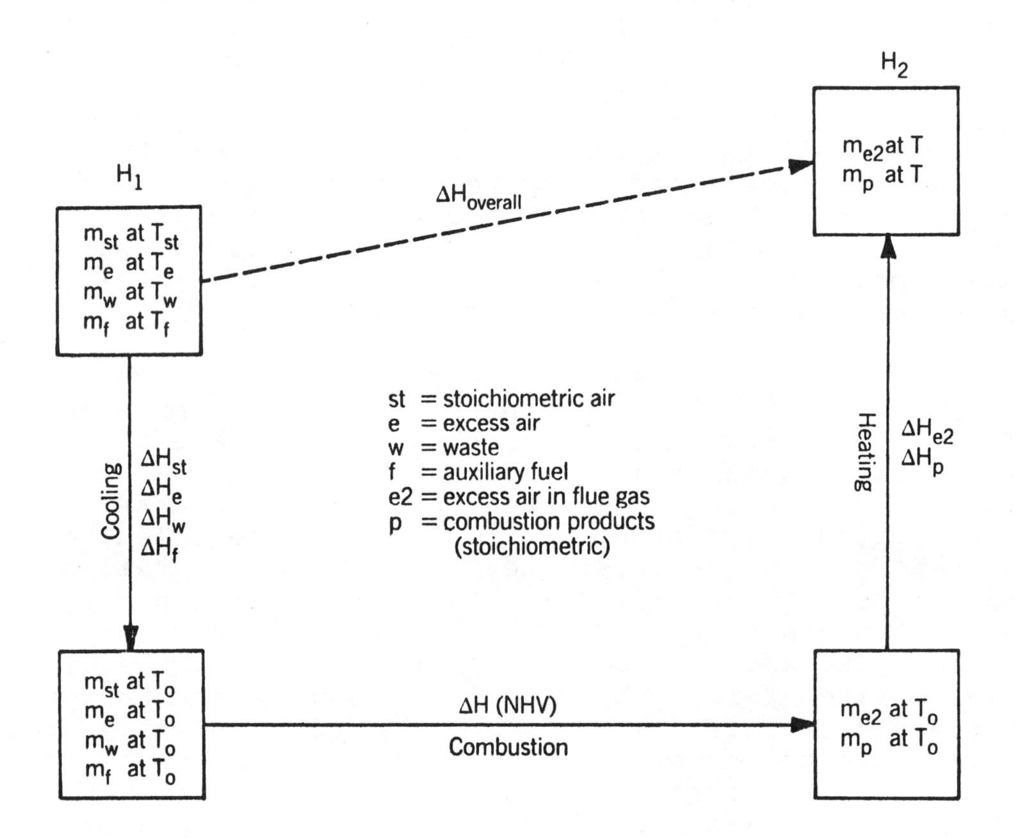

Figure (2-1). Flame temperature calculation schematic.

m_i = mass of *i* per unit mass of waste-fuel mixture, lb/lb mixture

c_{Pi} = average specific heat capacity of *i* over the temperature range T_i to T_0, Btu/lb-^oF

T_i = initial temperature of *i*, ^oF

T_0 = reference temperature, ^oF

subscripts St, E, W and F refer to the stoichiometric air, excess air, waste and fuel, respectively.

combustion step:

$$NHV = m_W\ (NHV_W) + m_F\ (NHV_F) \qquad (2\text{-}40)$$

where NHV = net heating value of waste-fuel mixture at T_0, Btu/lb mixture

NHV_W = net heating values of waste at T_0, Btu/lb waste

NHV_F = net heating value of feed at T_0, Btu/lb feed

heating step:

$$\Delta h_{E2} = m_E\ c_{PE2}\ (T - T_0)$$
$$\Delta h_P = \sum_P m_i\ c_{Pi}\ (T - T_0) \qquad (2\text{-}41)$$

where Δh_i = sensible enthalpy change for heating *i* from reference temperature T_0 to the incinerator operating temperature T, Btu/lb mixture

m_i = mass of *i* per unit mass of waste-fuel mixture, lb/lb mixture

c_{Pi} = average specific heat capacity of *i* over the temperature range T_0 to T

E2 = subscript indicating excess air in the flue gas

P = products of combustion (CO_2, H_2O, HCl, SO_2, N_2, etc.).

overall enthalpy balance:

$$h_2 - h_1 = \Delta h_{St} + \Delta h_E + \Delta h_W + \Delta h_F + (1-R)NHV + \Delta h_{E2} + \Delta h_P$$
$$= (R)NHV \qquad (2\text{-}42)$$

The term R refers to the fraction (not percent) of the net heating value associated with the combustion step that is lost by heat across the walls of the incinerator. If the system is assumed to be adiabatic, R is equal to zero. (Note: The *HWI* program automatically sets R to zero unless a value is supplied by the user.)

Equating the cooling and combustion enthalpy changes above with the heating and overall enthalpy changes provides a rigorous equation relating the incinerator operating temperature, T, with the system's variables.

Fortunately, a simpler algorithm[5] is available to perform these detailed thermodynamic calculations. This is presented in Equations (2-43) to (2-45).

$$T = 60 + (NHV)/[(0.3)[1 + (1 + EA)(7.5 \times 10^{-4})\ (NHV)] \tag{2-43}$$

This equation may be rearranged and solved in terms of the excess air (EA) and the net heating value of the waste-fuel mixture (NHV).

$$EA = \frac{[(NHV)/(0.3)(T-60)]\ -1}{(7.5 \times 10^{-4})(NHV)} - 1 \tag{2-44}$$

$$NHV = (0.3)(T-60)/[1 - (1 + EA)(7.5 \times 10^{-4})(0.3)(T - 60)] \tag{2-45}$$

Note that the units of T, EA, and NHV are ^{o}F, fractional basis and Btu/lb, respectively. The reader should also note that the above equations become sensitive to the value assigned to c_p (0.3 in the above case). This is a function of both the temperature (T) and excess air fraction (EA), and also depends on the flue products since the heat capacities of air and CO_2 are about half that of H_2O. These variabilities can be compensated for by treating the c_p term as a statistical parameter in Equations (2-43) through (2-45). Theodore and Reynolds[5] have generated values for c_p for excess air percent (EA) and operating temperatures (T) ranging from 0 to 100% and 1000 to 2750 K, respectively, for benzene, octane, and dichloromethane. In addition, regression coefficients α, β, and γ (not to be confused with the heat capacity coefficients) were obtained from these results for the equation

$$c_p = \alpha + \beta T + \gamma(EA) \tag{2-46}$$

These are found in Table (2-7).

Table (2-7). Regression coefficients for use with Eqn. (2-46)
Temperature range: 1000 - 2750 K
Excess Air range: 0 - 100%

Substance	Formula	α	$\beta \times 10^5$	$\gamma \times 10^3$
Benzene	C_6H_6	0.336	2.230	-0.942
Octane	C_8H_{18}	0.372	2.434	-1.311
Dichloromethane	CH_2Cl_2	0.233	2.645	-0.102
Benzene and Octane		0.354	2.332	-1.126
Overall		0.314	2.436	-0.785

CHEMICAL REACTION EQUILIBRIUM

With regard to chemical reactions, there are two important questions which are of concern to the engineer: (1) how *far* will the reaction go; and (2) how *fast* will the reaction go? Chemical thermodynamics provides the answer to the first question; however, it tells nothing about the second. Reaction rates fall within the domain of chemical kinetics and are treated elsewhere in the literature[1,4]. To illustrate the difference and importance of both questions in an engineering analysis of a chemical reaction, consider the following process. Substance A, which costs one cent per ton, can be converted to B, which costs one million dollars per pound, by the reaction A <-----> B. Chemical thermodynamics will provide information on the maximum amount of B that can be formed. If 99.99% of A can be converted to B, the reaction would then appear to be economically feasible, from a thermodynamic point of view. However, a kinetic analysis might indicate that the reaction is so slow that, for all practical purposes, its rate is vanishingly small. For example, it might take 10^6 years to obtain a 10^{-16}% conversion of A. The reaction is then economically unfeasible. Thus, it can be seen that both equilibrium and kinetic effects must be considered in an overall engineering analysis of a chemical reaction. The same principle applies to control of gaseous pollutants by combustion and incineration of hazardous wastes.

A rigorous, detailed presentation of this topic is beyond the scope of this text. However, a superficial treatment is presented in the hope that it may provide at least a qualitative introduction to chemical reaction equilibrium. This material does find application in incinerator calculations involving SO_3 and Cl_2 discharges and will hopefully explain, in part, the role of reaction equilibria in these calculations.

Chemical reaction equilibrium calculations are structured around another thermodynamic term referred to as *free energy*. This so-called free energy, G, is a thermodynamic property that cannot be easily defined without some basic grounding in thermodynamics. No attempt will be made to define it here and the interested reader is again directed to the literature[3] for further development of this topic.

Consider the equilibrium reaction

$$aA + bB = cC + dD \tag{2-47}$$

where A, B, C, D = chemical formulas of the reactant and and product species
a, b, c, d = stoichiometric coefficients

and the = sign is a reminder that the reacting system is at equilibrium. For the above reaction (as with enthalpy),

$$\Delta G^o = cG^o_C + dG^o_D - aG^o_A - bG^o_B \tag{2-48}$$

ΔG^o represents the free energy change for the above reaction when reactants and products are in their standard states. Note that ΔG^o may be calculated and obtained in a manner similar to ΔH^o. Note further that, at 25°C, the free energy of an element in its standard state is arbitrarily set at zero. Therefore

$$(\Delta G_f^o)_{i,298} = G^o_{i,\ 298}$$

so that

$$\Delta G^o_{298} = c(\Delta G_f^o)_C + d(\Delta G_f^o)_D - a(\Delta G_f^o)_A - b(\Delta G_f^o)_B \tag{2-49}$$

According to Equation (2-49), at 25°C, the standard free energy of reaction ΔG^o may be calculated from standard free energy of formation data. Some of this information is presented in Table (2-8).

Equation (2-50) is used to calculate the chemical reaction equilibrium constant K at a temperature T. This will be described in terms of physically measureable properties later in this section.

$$\Delta G^o_T = -RT \ln K \tag{2-50}$$

The value of this equilibrium constant depends on the temperature at which the equilibrium is established. The effect of temperature on K is now examined.

Effect of Temperature on the Equilibrium Constant

The dependence of ΔG^o on temperature is given by

Table (2-8). Standard free energy of formation at 25°C in cal/gmol[a]

Compound	Formula	State	ΔG°_{f298}
	Normal Paraffins		
Methane	CH_4	g	−12,140
Ethane	C_2H_6	g	−7,860
Propane	C_3H_8	g	−5,614
n-Butane	C_4H_{10}	g	−4,100
n-Pentane	C_5H_{12}	g	−2,000
n-Hexane	C_6H_{14}	g	−70
n-Heptane	C_7H_{16}	g	1,920
n-Octane	C_8H_{18}	g	3,920
Increment per C atom above C_8		g	2,010
	Normal Monoolefins (1-alkenes)		
Ethylene	C_2H_4	g	16,282
Propylene	C_3H_6	g	14,990
1-Butene	C_4H_8	g	17,090
1-Pentene	C_5H_{10}	g	18,960
1-Hexene	C_6H_{12}	g	20,940
Increment per C atom above C_6		g	2,010
	Miscellaneous Organic Compounds		
Acetaldehyde	C_2H_4O	g	−31,960
Acetic acid	$C_2H_4O_2$	l	−93,800
Acetylene	C_2H_2	g	50,000
Benzene	C_6H_6	g	30,989
Benzene	C_6H_6	l	29,756
1,3-Butadiene	C_4H_6	g	36,010
Cyclohexane	C_6H_{12}	g	7,590
Cyclohexane	C_6H_{12}	l	6,370
Ethanol	C_2H_6O	g	−40,130
Ethanol	C_2H_6O	l	−41,650
Ethylbenzene	C_8H_{10}	g	31,208
Ethylene glycol	$C_2H_6O_2$	l	−77,120
Ethylene oxide	C_2H_4O	g	−2,790
Methanol	CH_4O	g	−38,810
Methanol	CH_4O	l	−39,850
Methylcyclohexane	C_6H_{14}	g	6,520
Methylcyclohexane	C_6H_{14}	l	4,860
Styrene	C_8H_8	g	51,100
Toluene	C_7H_8	g	29,228
Toluene	C_7H_8	l	27,282
	Miscellaneous Inorganic Compounds		
Ammonia	NH_3	g	−3,976
Ammonia	NH_3	aq	−6,370
Calcium carbide	CaC_2	s	−16,200
Calcium carbonate	$CaCO_3$	s	−269,780
Calcium chloride	$CaCl_2$	s	−179,300
Calcium chloride	$CaCl_2$	aq	−194,880
Calcium hydroxide	$Ca(OH)_2$	s	−214,330
Calcium hydroxide	$Ca(OH)_2$	aq	−207,370
Calcium oxide	CaO	s	−144,400

Table (2-8). *(Continued)*

Compound	Formula	State	ΔG°_{f298}
Carbon dioxide	CO_2	g	−94,258
Carbon monoxide	CO	g	−32,781
Hydrochloric acid	HCl	g	−22,778
Hydrogen sulfide	H_2S	g	−7,892
Iron oxide	Fe_3O_4	s	−242,400
Iron oxide	Fe_2O_3	s	−177,100
Iron sulfide	FeS_2	s	−39,840
Nitric acid	HNO_3	l	−19,100
Nitric acid	HNO_3	aq	−26,410
Nitrogen oxides	NO	g	20,690
	NO_2	g	12,265
	N_2O	g	24,933
	N_2O_4	g	23,395
Sodium carbonate	Na_2CO_3	s	−250,400
Sodium chloride	$NaCl$	s	−91,785
Sodium chloride	$NaCl$	aq	−93,939
Sodium hydroxide	$NaOH$	s	−90,600
Sodium hydroxide	$NaOH$	aq	−100,184
Sulfur dioxide	SO_2	g	−71,790
Sulfur trioxide	SO_3	g	−88,520
Sulfuric acid	H_2SO_4	aq	−177,340
Water	H_2O	g	−54,635
Water	H_2O	l	−56,690
	Chlorinated Organics		
Methyl chloride	CH_3Cl	l	−15,030
Dichloromethane	CH_2Cl_2	l	−16,460
Chloroform	$CHCl_3$	l	−16,380
Carbon tetrachloride	CCl_4	l	−13,920
Ethyl chloride	C_2H_5Cl	l	−14,340
1,1-Dichloroethane	$C_2H_4Cl_2$	l	−17,470
1,1,2,2-Tetrachloroethane	$C_2H_2Cl_4$	l	−20,480
n-Propyl chloride	C_3H_7Cl	l	−12,110
1,3-Dichloropropane	$C_3H_6Cl_2$	l	−19,740
n-Butyl chloride	C_4H_9Cl	l	−9,270
1-Chloropentane	$C_5H_{11}Cl$	l	−8,940
1-Chloroethylene	C_2H_3Cl	l	−12,310
trans-1,2-Dichloroethylene	$C_2H_6Cl_2$	l	−6,350
Trichloroethylene	C_2HCl_3	l	−4,750
Tetrachloroethylene	C_2Cl_4	l	−4,900
3-Chloro-1-propene	C_3H_5Cl	l	10,420
Chlorobenzene	C_6H_5Cl	l	23,700
p-Dichlorobenzene	$C_6H_5Cl_2$	l	18,440
Hexachlorobenzene	C_6Cl_6	l	10,560

[a] Selected mainly from F.D. Rossini, ed., "Selected Values of Physical and Thermodynamic Properties of Hydrocarbons and Related Compounds," *American Petroleum Institute Research Project 44*, Carnegie Institute of Technology, Pittsburgh, PA, 1953; F.D. Rossini, D.D. Wagman, W.H. Evans, S. Levine, and I. Jaffe, "Selected Values of Chemical Thermodynamic Properties," *Natl. Bur. Stand. Circ.* **500**, 1952; and also from personal notes, L. Theodore and J. Reynolds, 1986.

$$\frac{d(\Delta G^o/RT)}{dT} = \frac{-\Delta H^o}{RT^2} \qquad (2-51)$$

Substitution of Eqn. (2-50) into Eqn. (2-51) yields

$$\frac{d\ (\ln K)}{dT} = \frac{\Delta H^o}{RT^2} \qquad (2-52)$$

Equation (2-52) shows the effect of temperature on the equilibrium constant, and hence on the equilibrium yield. If the reaction is exothermic, H^o is negative and the equilibrium constant decreases with an increase in temperature; for an endothermic reaction, the equilibrium constant increases with an increase in temperature.

If the term ΔH^o, which is the standard enthalpy change (heat of reaction), is assumed to be constant with temperature, Equation (2-52) can be integrated between the temperatures T and T_1 to give

$$\ln\ (K/K_1) = -\ (\Delta H^o/R)[(1/T) - (1/T_1)] \qquad (2-53)$$

This approximate equation may be used to determine the equilibrium constant at a temperature T from a known value at some other temperature T_1 if the difference between the two temperatures is small. If the standard heat of reaction is known as a function of temperature, however, Equation (2-52) can be integrated rigorously.

$$\ln K = (1/R) \int (\Delta H^o/T^2)dT + I \qquad (2-54)$$

where I = constant of integration.

In an earlier section, it was shown that, if the molar heat capacities for each chemical species taking part in the reaction are known and can be expressed as a power series in T (degrees K)

$$C_p = a + bT + cT^{-2}$$

then ΔH^o at a given temperature T becomes

$$\Delta H^o{}_T = \Delta H_o + \Delta a\ T + (\Delta b/2)T^2 - \Delta cT^{-1}$$

In this equation, the constant ΔH_o can be calculated from the standard heat of reaction at 25°C (see Eqn. 2-30). With ΔH_o determined, $\Delta H^o{}_T$ can be substituted into Equation (2-54) and integrated to obtain

$$\ln K = -(\Delta H^{o}/RT) + (\Delta a/R)\ln T + (\Delta b/2R)T + (\Delta c/2R)T^{-2} + I \quad (2\text{-}55)$$

Here the constant I may be evaluated from a knowledge of the equilibrium constant at one temperature. This is usually obtained from standard free energy of formation data at 25°C. Equation (2-50) is employed to obtain K at this temperature. A similar equation for G^{o}_{T} may be obtained by combining Equations (2-50) and (2-55).

$$\Delta G^{o}_{T} = \Delta H_{o} - (\Delta a)T \ln T - (\Delta b/2)T^{2} - (\Delta c/2)T^{-1} - IRT \quad (2\text{-}56)$$

The reader is left the exercise of developing companion equations to Equations (2-54) and (2-56) if the heat capacity variation with temperature is given by

$$C_{p} = \alpha + \beta T + \gamma T^{2}$$

Equilibrium Conversion

The problem that remains is to relate K to measureable physical quantities. For gas phase reactions, as in an incinerator operation, the term K for Equation (2-42) may be approximately represented in terms of the partial pressures (in atmospheres) of the components involved. This functional relationship is given in Equation (2-57).

$$K = K_{P} \quad (2\text{-}57)$$

where K = equilibrium constant

$$K_{P} = P(C)^{c}\, P(D)^{d} \,/\, P(A)^{a}\, P(B)^{b} \quad (2\text{-}58)$$

P(i) = partial pressure of component *i*, atm

This definition of K_{P} obviously applies to the reaction of Equation (2-47). Assuming a K value is available or calculable, this equation may be used to determine the partial pressures of the participating components at equilibrium. It is important to note that the component partial pressures, P(i), are equilibrium values. For product gases, P(i) usually represents the maximum values that can ultimately be achieved; for reactant gases, P(i) represents minimum values.

Theodore and Reynolds[5] have obtained the chemical reaction equilibrium constant based on partial pressures (K_{P}) as a function of temperature for two key incinerator reactions. For the reaction

$$2\ HCl + 0.5\ O_2 = Cl_2 + H_2O$$

the final result takes the form (at 1.0 atm):

$$\ln K = 7048.7/T + 0.0151(\ln T) - 9.06\times10^{-5}(T) - 2.714\times10^{4}(T^{-2}) - 8.09 \quad (2\text{-}59)$$

From Equation (2-58),

$$K = K_P = \frac{P(Cl_2)\ P(H_2O)}{P(HCl)^2\ P(O_2)^{1/2}} \quad (2\text{-}60)$$

K_P for the reaction (at 1 atm pressure)

$$SO_2 + 0.5\ O_2 = SO_3$$

may be expressed in the form:

$$\ln K = 11{,}996/T - 0.362(\ln T) + 9.36\times10^{-4}(T) - 2.969\times10^{5}(T^{-2}) - 9.88 \quad (2\text{-}61)$$

Applying Equation (2-58) to this reaction,

$$K = K_P = \frac{P(SO_3)}{P(SO_2)\ P(O_2)^{1/2}} \quad (2\text{-}62)$$

INCINERATOR DESIGN

The physical design of an incinerator is primarily dictated by the allowable heat release rate, q_H, with units of Btu/hr-ft^3. The volume, V, of the incinerator is then given by

$$V = q/q_H \quad (2\text{-}63)$$

where q is the rated capacity of the incinerator in Btu/hr. The length and diameter of the unit is obtained from the length-to-diameter ratio (L/D). Additional details on this calculation are given in Chapter 3.

CHAPTER THREE

Program Details

This chapter will serve to introduce the reader to some of the details associated with the *HWI* (Hazardous Waste Incineration) computer program that performs the key engineering calculations for a hazardous waste incinerator. The reader is referred to the User's Guide (Part 3 of this book) for more complete and extensive details.

In Chapter 2, a number of scientific principles, mostly from the fields of chemistry and chemical engineering, were presented. The presentation applies to almost any type of chemical reaction. In Chapter 3, these same concepts are applied directly to the incineration process. The conservation laws and stoichiometry principles are employed in the stoichiometric calculations. This involves the prediction of the flue gas composition and flow rate from the incineration of waste-fuel mixtures of known composition. The thermochemistry and chemical reaction equilibrium principles from Chapter 2 are also applied here to perform incinerator thermochemical calculations. These are used to predict the incineration operating temperature as a function of waste-fuel mixture, net heating value and excess air ratio.

As is the case with most computer programs, *HWI* is organized into three parts: input of data, calculations, and output of results. The *HWI* output is demonstrated in Chapters 4 and 5 of Part 4. Input of data and calculational details are covered in separate sections below.

INPUT OF DATA

The inputting of the data required for the calculations is accomplished at the computer terminal and is handled in question-and-answer fashion. Prompts appear on the monitor screen and the user answers them from the keyboard. In many cases, a number of options for inputting a particular piece of information is provided.

The required data to be inputted by the user include: calculational mode, waste composition, waste flow rate, air requirements, waste heating value, heat loss from the incinerator, and some design information.

1. <u>Calculational mode</u>. The three key parameters in the *HWI* calculations are the incinerator operating temperature, the waste-fuel heating value and the amount of excess air. The waste-fuel heating value is determined by the materials being

combusted. Either one of the remaining two parameters may be treated as the independent variable. The *HWI* program provides two options: (1) the amount of excess air to be employed may be determined by the user, and the resulting incinerator operating temperature calculated by *HWI*; or (2) the desired operating temperature may be inputted by the user and the required amount of excess air calculated by the program.

2. Waste composition. In order to accommodate various methods of characterizing the chemical make-up of the waste, *HWI* provides the user with the following options:

1 - elemental (or ultimate analysis), atom basis
2 - elemental analysis, mass basis
3 - componential analysis, mole basis
4 - componential analysis, mass basis

In a *componential* analysis, the composition of the waste (or, when auxiliary fuel has been added, the waste-fuel mixture) is described in terms of what chemical *compounds* are present and the fraction or percent of each. In an *elemental* analysis, the composition is given in terms of what *elements* are present and the fraction or percent of each. Whichever of these four options is employed by the user during the input phase of the program, the composition is converted to an elemental analysis on a mole fraction basis before the stoichiometric calculations are begun.

If one of the elemental analysis options is chosen (Option 1 or 2), a prompt appears on the screen for the number of elements. For each element, the user is then requested to input the element's chemical symbol and its fractional or percentage contribution to the waste (atom percent or fraction for Option 1, mass percent or fraction for Option 2). If one of the componential analysis options is chosen (Option 3 or 4), a prompt appears for the number of compounds. For each compound, the user is requested to input the percent or fraction of that compound (mole or mass percent for Option 3 or 4, respectively) and the number of elements in that compound. For each element in the compound, the user is asked for the chemical symbol and the number of atoms per molecule. Thus, in effect, the user types in the compound's chemical formula.

The following elements are "recognized" or accepted by *HWI*: carbon, hydrogen, oxygen, nitrogen, sulfur, chlorine, fluorine, bromine and iodine. Inert material (ash) which can be assumed to exit the incinerator as solids (particulates) entrained in the flue gas stream may also be inputted as part of the waste-fuel composition. This material is assumed to have negligble impact on the incinerator processes.

3. Waste flow rate. The waste (or waste-fuel mixture) flow rate is the next input requested. With Composition Option 3 or 4, the user has the choice of lb/hr or lbmol/hr; for Option 1 or 2, lb/hr must be used.

4. Air requirements. If Calculational Mode 1 is being used (i.e., the user inputs the amount of excess air and *HWI* is to calculate the operating temperature), the amount of excess air is inputted at this point. If Calculational Mode 2 is being used (i.e., the user inputs the desired operating temperature and HWI is to calculate the required amount of excess air), only the temperature of the preheated air is called for (see below).

Since the amount of excess air may depend on the amount of stoichiometric or theoretical air (i.e., the exact amount of air required to convert each element in the waste to its combustion product, assuming complete combustion), this quantity is calculated and displayed on the screen prior to the prompt for the percent of excess air. Details of this calculation will be presented in the next section (Calculations). If the air is preheated (for example, by means of a heat exchanger), the resulting incinerator temperature will be higher for the same amount of input air. The next prompt asks for the temperature of the preheated air.

5. Waste heating value. The net heating value of the waste may either be inputted by the user or estimated by Dulong's method (Equation 1). If NHV data are available, and if one of the componential options was used to input the waste composition, the user has a choice of inputting the NHVs of each component or the NHV of the waste as a whole. If one of the elemental options was used, only the overall NHV may be inputted. The user is given the choice of several different units for the NHV input(s).

6. Heat loss. In practice, any heat loss from the incinerator is often neglected. If the heat loss is to be taken into account, a prompt is provided for its input in terms of percent of the NHV.

7. Preliminary design data. The last set of input data is required for a preliminary design (sizing) of the incinerator. For this purpose, either the superficial throughput velocity of the flue gas or the length-to-diameter ratio (L/D) of the incinerator must be specified by the user. If the user omits this input, a L/D of 3.0 is assumed. The heat release rate, which is a function of the type of incinerator and, to a lesser extent, the nature of the waste, is another required design parameter. If the user ignores this input, a release rate of 25,000 Btu/hr-ft^3 (a fairly typical value) is assumed.

CALCULATIONS

After the last input, the computer performs the stoichiometric and thermodynamic calculations, a process which takes only a few seconds. First, the composition of the flue gas is determined. The flue gas consists of the products of combustion of the waste plus nitrogen and any excess oxygen from the air. The combustion reactions for each of the elements in the waste are assumed to be:

$$C \longrightarrow CO_2$$

$$H \longrightarrow H_2O$$

$$Cl \longrightarrow HCl$$

$$O \longrightarrow O_2$$

$$N \longrightarrow N_2$$

$$S \longrightarrow SO_2$$

$$F \longrightarrow HF$$

$$Br \longrightarrow Br_2$$

$$I \longrightarrow I_2 \qquad (3\text{-}1)$$

A balanced stoichiometric equation of the following form is employed.

$$C_zH_yO_xCl_wS_vN_uF_rBr_sI_t + [z + \phi + v - 0.5x]\ O_2 + (79/21)\ (z + \phi + v - 0.5x)]\ N_2$$

$$\longrightarrow z\ CO_2 + 2\phi\ H_2O + w\ HCl + v\ SO_2 + r\ HF + 0.5s\ Br_2 + 0.5t\ I_2 + [0.5u + (79/21)\ (z + \phi + v - 0.5x)]\ N_2 \qquad (3\text{-}2)$$

where z,y,x,u,v,w,r,s,t = number of moles (or mole fractions) of C,H,O,Cl,S,N,F,Br,I present in the waste-fuel mixture, respectively

$$\phi = (1/4)\ (y-w-r),\ \text{when } y > (w+r)$$

$$\phi = 0\ ,\ \text{when } y \leqslant (w+r)$$

Note that the parameter ϕ is included if its value is positive, and ignored if it is zero or negative. For the above approach, the mole fraction for carbon in the waste-fuel mixture, x_C, is given by

$$x_C = z / (z + y + x + u + v + w + r + s + t) \qquad (3\text{-}3)$$

with similar equations applicable to hydrogen, oxygen, etc. This assumes that the waste-fuel mixture consists only of C,H, O, N, S, Cl, F, Br and I. Equations (3-2) and (3-3) are valid provided:

1. Cl_2, SO_3 and NO_x formations are neglected (for the time being).

2. Oxygen in the waste-fuel mixture is available for combustion and is therefore treated as a credit in calculating stoichiometric air requirements.

3. The air essentially consists of 79% nitrogen and 21% oxygen (by mole or volume).

4. The combustion reactions go to completion.

Note that the second and third terms on the left-hand side of Equation (3-2) represent the stoichiometric air only. The excess air is taken into account as follows. If EA is the fraction of excess air, then the factor **(79/21)** in the coefficients of N_2 on both sides of the equation should be replaced by **(1+EA)(79/21)**; and the term, **(EA)**O_2, should be added on the right.

The calculation of the flue gas composition by the stoichiometric method shown above is not difficult. However, the hydrogen chloride and sulfur dioxide are not completely stable and, as shown earlier, are involved in equilibrium reactions (cf. Equation 2-59). This makes the calculations much more complicated because the equilibrium concentrations of chlorine and sulfur trioxide depend on the incinerator temperature (cf. Equation 2-60). The temperature reached in the incinerator, however, depends to some extent on the flue gas composition, because almost all of the energy released by the incineration reactions is absorbed by the flue gas, the heat capacity of which depends on its composition.

A further complication results from the fact that the two equilibrium equations are interdependent since both involve the concentration of the oxygen. This is shown in Equations (2-60) and (2-62); note that the partial pressure of oxygen appears in both.

The incinerator operating temperature is calculated by the enthalpy balance shown in Equation (2-42). However, the last two terms of that equation cannot be determined exactly until the flue gas composition is known. Since flue gas composition is a function of temperature, an initial estimate or guess of the temperature must be made. Equation (2-43), which is based on a simplified enthalpy balance and an assumed average flue gas heat capacity of 0.3 Btu/lb-^{o}F, is employed for this purpose. Using this estimated temperature, the two equilibrium constants are calculated from Equations (2-59) and (2-61). Numerical methods are then used to solve Eqns. (2-60) and (2-62) simultaneously and the flue gas composition is recalculated, this time including the sulfur trioxide and chlorine. A more accurate incinerator temperature is next calculated from the enthalpy balance, Eqn. (2-42), but this time, the ΔH_{E2} and ΔH_P terms are evaluated using the corrected flue gas composition. (Note that the heat capacities of the components of the flue gas are functions of temperature, and the calculations of ΔH_{E2} and ΔH_P involve an integration of the heat capacities from room temperature to the incinerator temperature.) New equilibrium constants and a new flue gas composition based on this revised temperature are next calculated. If necessary, this process or loop is repeated several times until the calculated temperature, equilibrium constants, and compositions converge to constant values.

The incinerator temperature and flue gas composition are used to determine the flue gas volumetric flow rate at actual (acfm), standard (scfm), and dry standard (dscfm) conditions, as well as the molar flow rate.

The heat release rate, q_H, is employed to determine the incinerator volume, V, as shown in Equation (2-63):

$$V = q/q_H$$

From the volume and either the length-to-diameter ratio, L/D, or superficial throughput velocity, v (whichever one was inputted by the user), the dimensions of the incinerator are calculated through the use of Equations (3-4) to (3-7). For a specified L/D ratio:

$$D = [4V/\pi(L/D)]^{1/3} \qquad (3\text{-}4)$$

For a specified velocity, the cross-sectional area, A, and diameter, D, are calculated from the volumetric flow rate, $\dot{V}$:

$$A = \dot{V}/v \qquad (3\text{-}5)$$

$$D = (4A/\pi)^{1/2} \qquad (3\text{-}6)$$

$$L = V/A \qquad (3\text{-}7)$$

The residence time, t, is calculated by

$$t = V/\dot{V} \qquad (3\text{-}8)$$

CHAPTER FOUR

Chlorobenzene/Sulfur Mixture Example

This manual contains two illustrative examples. The first example, presented here in chapter 4, is concerned with the incineration of a hazardous waste that contains -- on an elemental basis -- carbon, hydrogen, chlorine, and sulfur. The waste is a 95%/5% (by mass) chlorobenzene/sulfur mixture and is to be incinerated in a liquid injection unit. The approach described in Chapter 3 is employed in the solution and is essentially the method used in the *HWI* program. Several of the tables presented in this chapter were generated by the program.

Problem Statement

5000 lb/hr of a hazardous waste is to be incinerated in a liquid injection unit with 100% excess air. The waste has the following composition:

Element	Mass %
carbon	60.8
hydrogen	4.2
chlorine	30.0
sulfur	5.0

a. Calculate the flue gas composition and flow rate assuming complete combustion.

b. Calculate the incinerator temperature.

c. Determine the length and diameter of the incinerator. Assume a flue gas superficial throughput velocity of 20 ft/sec.

Problem Solution

Since heat of combustion data are not given and since the waste contains only carbon, hydrogen, chlorine and sulfur, Dulong's method (Eqn. II-31) is used to estimate the NHV.

$$NHV = 14000\ m_C + 45000\ (m_H - 0.125\ m_O) - 760\ m_{Cl} + 4500\ m_S$$

$$= 14000\ (0.608) + 45000\ (0.042) - 760\ (0.300) + 4500\ (0.050)$$

$$= 10400\ \text{Btu/lb} = 5777\ \text{cal/g} \qquad (4\text{-}1)$$

To obtain a first estimate of the average operating temperature of the incinerator, Equation (2-43) is applied.

$$T = 60 + (NHV/0.3)/[1 + (1+EA)(7.5x10^{-4})(NHV)]$$

$$= 60 + (10400/0.3)/[1 + (1+1.0)(7.5x10^{-4})(10400)]$$

$$= 2148^{o}F = 1449\ K \qquad (4-2)$$

Chemical reaction equilibrium constants for the HCl/Cl_2 and SO_2/SO_3 distributions at 1449 K are now obtained from Equations (2-59) and (2-60).

For HCl/Cl_2:

$$\ln K_1 = 7048.7/T + 0.0151 \ln T - 9.06x10^{-5}\ T - 2.714x10^4\ T^{-2} - 8.09$$

When T = 1449 K,

$$\ln K_1 = 7048.7/1449 + 0.0151 \ln(1449) - 9.06x10^{-5}(1449) - 2.714x10^4(1449)^{-2} - 8.09$$

$$= - 3.260$$

$$K_1 = 0.03839\ atm^{-1/2} \qquad (4-3)$$

For SO_2/SO_3:

$$\ln K_2 = 11996/T - 0.362 \ln T + 9.36x10^{-4}\ T - 2.969x10^5\ T^{-2} - 9.88$$

When T = 1449 K,

$$\ln K = 3.022$$

$$K_2 = 0.04872\ atm^{-1/2} \qquad (4-4)$$

The stoichiometric equations are written using Equation (3-2) and the initial compositions. [The fluorine, bromine and iodine components of Eqn. (3-2) have been ignored.]

$$C_zH_yO_xN_uS_vCl_w + [z + o + v - 0.5x]\ O_2 + (79/21)(z + \phi + v - 0.5x)\ N_2$$

$$\text{----->}\ z\ CO_2 + 2\phi\ H_2O + w\ HCl + v\ SO_2 + [0.5u + (79/21)(z + \phi + v - 0.5x)]\ N_2$$

where $\phi = (y - w)/4$ if $y > w$

The elemental analysis of the waste is converted to an atom basis, as shown in Table (4-1).

Table (4-1). Waste composition.

ELEMENT	MASS%	LB/HR	AW	LB-ATOM*	ATOM %
CARBON	60.80	3040.	12.01	5.062	49.48
HYDROGEN	4.20	210.	1.01	4.167	40.73
CHLORINE	30.00	1500.	35.45	0.846	8.27
SULFUR	5.00	250.	32.06	0.156	1.52
TOTAL	100.00	5000.		10.231	100.00

*Basis: 100 lb waste

When one lbmole of waste is used as a basis, the constants in Eqn. (4-5) are $z = 0.506$, $y = 0.417$, $x = 0.0$, $u = 0.0$, $v = 0.016$, and $w = 0.085$. Then $\phi = (0.417 - 0.085)/4 = 0.083$ and Eqn. (4-5) becomes

$$C_{0.506}\,H_{0.417}\,S_{0.016}\,Cl_{0.085} + 0.605\,O_2 + 2.276\,N_2$$
$$\text{----->}\ 0.506\,CO_2 + 0.166\,H_2O + 0.085\,HCl + 0.016\,SO_2$$
$$+ 2.276\,N_2 \qquad (4\text{-}6)$$

With 100% excess air, the oxygen and nitrogen in the stoichiometric equation are doubled. Therefore, Equation (4-6) becomes

$$C_{0.506}\,H_{0.417}\,S_{0.016}\,Cl_{0.085} + 1.210\,O_2 + 4.552\,N_2$$
$$\text{----->}\ 0.506\,CO_2 + 0.166\,H_2O + 0.085\,HCl + 0.016\,SO_2$$
$$+ 0.605\,O_2 + 4.552\,N_2 \qquad (4\text{-}7)$$

The HCl/Cl_2 and SO_2/SO_3 splits can now be calculated. The expressions for the equilibrium constants are given by Eqns. (2-60) and (2-62).

$$K_1 = P(Cl_2)\;P(H_2O)\;/\;P(HCl)^2\;P(O_2)^{1/2}$$

$$K_2 = P(SO_3)\;/\;P(SO_2)\;P(O_2)^{1/2}$$

The partial pressures of the components can be calculated using the coefficients in the reaction equation (4-7). The total number of moles in the flue gas per lbmole of waste mixture are

$$N_T = 0.506 + 0.166 + 0.085 + 0.016 + 0.605 + 4.552$$

= 5.930 lbmol

For an incinerator operating at 1.0 atm, the partial pressures are equal to mole fractions. Therefore,

$$P(CO_2) = 0.506/5.930 = 0.0854$$

The partial pressure results for CO_2 and the remaining components are provided in Table (4-2).

Table (4-2). Initial flue gas partial pressures.

Component	Partial pressure (atm)
CO_2	0.0854
H_2O	0.0280
HCl	0.0143
SO_2	0.00262
O_2	0.1020
N_2	0.7677
TOTAL	1.0000

When the equilibrium reactions

$$2\ HCl + 0.5\ O_2 = H_2O + Cl_2$$

and

$$SO_2 + 0.5\ O_2 = SO_3$$

are considered, the partial pressures are corrected to those shown in Table (4-3). In this table, z and y represent the number of moles of Cl_2 and SO_3 formed when equilibrium is achieved, and N is the total number of moles of flue gas at equilibrium starting with one mole of flue gas originally. Note that N is the sum of Column 2 and is equal to 1.0000-0.5z-0.5y.

Substituting these partial pressures and the values of the equilibrium constants found in Eqns. (4-2) and (4-3) into Eqns. (2-60) and (2-62) yields

$$0.03839 = (0.0280+z)(z)/(0.0143-2z)^2(0.1020-0.5z-0.5y)^{1/2}N^{1/2} \qquad (4\text{-}8)$$

Table (4-3). Corrected flue gas partial pressures.

Component	Number of Moles of Component*	Partial Pressure (atm)
CO_2	0.0854	(0.0854)/N
H_2O	0.0280 + 2	(0.0280 = z)/N
HCl	0.0143 - 2z	(0.0143 - 2z)/N
SO_2	0.00262 - y	(0.00262 - y)/N
O_2	0.1020-0.5z-0.5y	(0.1020-0.5z-0.5y)/N
N_2	0.7677	(0.7677)/N
Cl_2	z	z/N
SO_3	y	y/N
Total	1.0000-0.5z-0.5y	1.0000

Basis: 1 mole of flue gas before equilibrium reactions occur.

and

$$0.04872 = (y)/(0.00262-y)(0.1020-0.5z-0.5y)^{1/2}N^{1/2} \qquad (4-9)$$

where N = 1.0000-0.5z-0.5y.

Note that N, which appears in each partial pressure term, has been factored and appears only once in each of the two equations.

Eqns. (4-8) and (4-9) must be solved simultaneously to determine the values of z and y. The solution may be simplified by noting from the magnitudes of K_1 and K_2 that the equilibrium in each reaction heavily favors the reactants. In other words, the values of z and y are small in comparison to the original partial pressures. Eqns. (4-8) and (4-9) may be simplified to

$$0.03839 = (0.0280)(z)/(0.0143)^2(0.1020)^{1/2}(1.0000)^{1/2} \qquad (4-8a)$$

$$0.04872 = (y)/(0.00262)(0.1020)^{1/2}(1.0000)^{1/2} \qquad (4-9a)$$

Solutions of (4-8a) and (4-9a) yield values of z and y of 8.954×10^{-5} and 4.077×10^{-5} atm, respectively. These values may then be used to correct the partial pressures in Eqns. (4-8) and (4-9) and calculate more accurate values of z and y.

$$0.03839 = (0.0281)(z)/(0.0141)^2(0.1019)^{1/2}(0.9999)^{1/2} \quad (4\text{-}8b)$$

$$0.04872 = (y)/(0.00258)(0.1019)^{1/2}(0.9999)^{1/2} \quad (4\text{-}9b)$$

Eqns. (4-8b) and (4-9b) yield values of z and y of 8.639×10^{-5} and 4.012×10^{-5} atm, respectively. Further refinement yields the corrected partial pressures shown in Table (4-4). The flue gas composition table, Table (4-5), is based on the partial pressures shown in Table (4-4).

Table (4-4). Corrected flue gas partial pressures.

Component	Partial pressure (atm)
CO_2	0.0854
H_2O	0.0281
HCl	0.0141
SO_2	0.00258
O_2	0.1020
N_2	0.7677
Cl_2	8.647×10^{-5}
SO_3	4.013×10^{-5}
TOTAL	1.0000

Note that Table (4-5) does not represent the final flue gas composition since these results are based on an approximate incinerator temperature of 2148°F. A more accurate temperature is now determined from an enthalpy balance. In order to calculate the enthalpy change that occurs when the flue gases are heated from the 60°F reference temperature to the adiabatic flame (incinerator) temperature, the heat capacity as a function of temperature is required. The data of Table (4-6) are taken from Table (2-4). The second column is based on one gram of the waste-fuel mixture and the units on the heat capacity is cal/gmol-K.

Table (4-5). Flue gas composition.

COMPONENT	LB*	MASS %	LBMOL*	MOLE %
CARBON DIOXIDE	222.87	12.65	5.064	8.54
WATER (g)	30.01	1.70	1.665	2.81
NITROGEN	1275.08	72.35	45.522	76.77
OXYGEN	193.52	10.98	6.047	10.20
HYDROGEN CHLORIDE	30.55	1.73	0.838	1.41
CHLORINE	0.31	0.02	0.004	0.01
SULFUR DIOXIDE	9.87	0.56	0.154	0.26
SULFUR TRIOXIDE	0.15	0.01	0.002	0.00
TOTAL	1762.35	100.00	59.297	100.00

*Basis: 100 lb waste-fuel mixture.

Table (4-6). Heat capacity vs. temperature data.*

Component	gmole per gram of waste	a	$bx10^3$	$cx10^{-5}$
CO_2	$5.064x10^{-2}$	10.57	2.10	-2.06
H_2O	$1.665x10^{-2}$	7.30	2.46	0.00
HCl	$0.838x10^{-2}$	6.27	1.24	0.30
SO_2	$0.154x10^{-2}$	11.04	1.88	-1.84
O_2	$6.047x10^{-2}$	7.16	1.00	-0.40
N_2	$45.522x10^{-2}$	6.83	0.90	-0.12
Cl_2	$0.004x10^{-2}$	8.85	0.16	-1.80
SO_3	$0.002x10^{-2}$	13.90	6.10	-3.22

*$C_p = a + bT + cT^{-2}$ (cal/gmol-K)

If, by definition,

$$\Delta a = \sum_i (gmol_i)(a_i), \text{ etc....}$$

then, for this system,

$\Delta a = 4.269$

$\Delta b = 6.309\text{x}10^{-4}$

$\Delta c = -1.836\text{x}10^{4}$

If the waste, auxiliary fuel (when used) and air are assumed to be close to room temperature when fed into the incinerator and if the combustion is assumed to be adiabatic, Eqn. (2-42) reduces to

$$0 = NHV + \Delta H_{FG} \quad (2\text{-}42a)$$

where $\Delta H_{FG} = \Delta H_{E2} + \Delta H_{P}$

Equation (2-12) is used to evaluate ΔH_{FG} with $T_0 = 298$ and $T_1 = T$ (incinerator temperature).

$$0 = (-5777) + 4.269\ (T-298) + (0.5)(6.309\text{x}10^{-4})(T^2 - 298^2)$$
$$- (-1.836\text{x}10^{4})[(1/T) - (1/298)] \quad (4\text{-}10)$$

Note that, for thermodynamic consistency, if the ΔH_{FG} term in Eqn. (2-42a) is considered positive because the system (waste products) is gaining energy as the temperature rises from 298°K to T, the NHV term must be negative (-5777 cal/gmol) because the system is releasing energy during the combustion step.

Solution of (4-10) by trial-and-error yields

$T = 1503\text{ K} = 2245^{\circ}\text{F}$

The equilibrium constants must be recalculated at this revised temperature from Eqns. (2-59) and (2-60). The results are

$K_1 = 0.03213\ \text{atm}^{-1/2}$

$K_2 = 0.03796\ \text{atm}^{-1/2}$

The partial pressures of the eight flue gas components are recalculated using the more accurate equilibrium constants. The final flue gas composition is shown in Table (4-7). Note that, for greater accuracy, the process just described could be repeated. The results in Table (4-7) could be used with a new enthalpy balance to calculate a new incinerator temperature; revised values of K_1 and K_2 based on this new temperature could be determined; and a new composition table generated. The *HWI* program does this, although the results would not differ much from those shown in Table (4-7).

Table (4-7). Final flue gas composition.

COMPONENT	LB*	MASS %	LBMOL*	MOLE %
CARBON DIOXIDE	222.87	12.65	5.064	8.54
WATER (g)	30.01	1.70	1.665	2.81
NITROGEN	1275.08	72.35	45.522	76.77
OXYGEN	193.52	10.98	6.047	10.20
HYDROGEN CHLORIDE	30.55	1.73	0.838	1.41
CHLORINE	0.31	0.02	0.004	0.01
SULFUR DIOXIDE	9.87	0.56	0.154	0.26
SULFUR TRIOXIDE	0.15	0.01	0.002	0.00
TOTAL	1762.35	100.00	59.297	100.00

*Basis: 100 lb waste-fuel mixture.

The mass and molar flow rates of the flue gas ($\dot{m}$ and $\dot{n}$, respectively) are calculated from the results in Table (4-7).

$$\dot{m} = (1762.35 \text{ lb}/100 \text{ lb}_w)(5000 \text{ lb}_w/\text{hr}) = 88{,}120 \text{ lb/hr} \quad (4\text{-}11)$$

$$\dot{n} = (59.297 \text{ lbmol}/100 \text{ lb}_w)(5000 \text{ lb}_w/\text{hr}) = 2{,}965 \text{ lbmol/hr} \quad (4\text{-}12)$$

The volumetric flow rate (V) is calculated using the Ideal Gas Law (P = 1 atm).

$$\dot{V} = \dot{n}RT/P = (2965)(0.7302)(2245+460)/(1)$$

$$= 5.8570\text{x}10^6 \text{ ft}^3/\text{hr} = 97{,}617 \text{ acfm} \quad (4\text{-}13)$$

The volume of the incinerator is calculated from Eqn. (3-8) using a heat release rate (q_H) of 25,000 Btu/hr-ft^3 (since no value was specified). The heat rate generated by the combustion at 60°F (q) is

$$q = (\text{NHV})(\dot{m}_w) = 10400\ (5000) = 5.200\text{x}10^7 \text{ Btu/hr} \quad (4\text{-}14)$$

Then

$$V = q/q_H = 5.200\text{x}10^7/25{,}000 = 2080 \text{ ft}^3 \quad (4\text{-}15)$$

From the problem statement, the approximate physical dimensions of the unit are to be based on a superficial velocity of 20 ft/sec.

$$A = \dot{V}/v = 97{,}617/(20)(60) = 81.35 \text{ ft}^2 \quad (4\text{-}16)$$

Since $A = \pi D^2/4$,

$$D = 10.18 \text{ ft} \quad (4\text{-}17)$$

The length of the incinerator is given by

$$L = V/A = 2080/81.35 = 25.57 \text{ ft} \quad (4\text{-}18)$$

These dimensions yield a L/D ratio of 2.51, which is a reasonable value.

The overall residence time is calculated using Equation (3-10).

$$t = V/\dot{V} = (2080/97{,}620)60 = 1.28 \text{ sec} \quad (4\text{-}19)$$

This is a reasonable value. In general, a residence time below 0.75 sec is unacceptable.

Some final results for this problem are presented in Tables (4-8) and (4-9) which were generated by the *HWI* program.

Table (4-8). Flue gas data.

Mass flow rate, lb/hr	88118.
Volumetric flow rate, acfm	97617.
Volumetric flow rate, scfm (60 $^\circ$F)	18763.
Volumetric flow rate, dscfm (60 $^\circ$F)	18236.
Molar flow rate, lbmol/hr	2965.
Temperature, $^\circ$F	2245.
Sulfur trioxide concentration, ppm	31.5
Chlorine concentration, ppm	72.9
HCl mass flow rate, lb/hr	1527.4
Particulate loading, gr/acf	0.00
Particulate loading, gr/dscf -- corrected to 50% excess air	0.00

Table (4-9). Preliminary design data.

Heat generation rate, Btu/hr	0.5199E+08
Volume, cu ft	2080.
Diameter, ft	10.2
Length, ft	25.6
L/D ratio	2.51
Superficial velocity, ft/sec	20.00
Residence time, sec	1.28

CHAPTER FIVE

Chlorobenzene/DDT/Water Mixture Example

This manual concludes with an illustrative example concerned with the incineration of a hazardous waste mixture. The problem statement is given below. This in turn is followed by the solution that, for the most part, follows the calculated sequence provided in Chapter 3, Part 4. (The reader is referred to the *User's Guide* (Part 3 of this book) for additional calculational details).

Problem Statement

A hazardous waste has the following composition:

Component	Mass %
Chlorobenzene, C_6H_5Cl	58
DDT, $C_{14}H_9Cl_5$	26
Water, H_2O	16

Calculate the flue gas flow rate and composition when 6430 lb/hr of the waste is incinerated in a rotary kiln incinerator with 75% excess air. Also calculate the operating temperature and dimensions of the cylindrical housing. Assume an overall L/D ratio of 4.0 for the kiln/afterburner unit.

From heat of formation data, the following values for the heat combustion (NHV) of the three components can be calculated. These are:

Chlorobenzene	714,361 cal/gmol
DDT	1,580,000 cal/gmol
Water	-10,519 cal/gmol

(The "heat of combustion" listed for water is actually the heat of vaporization. In the incinerator, the water simply vaporizes.)

Problem Solution

The waste composition data are first converted to mole percents. The results are given in Table (5-1).

Table (5-1). Waste componential composition.

COMPONENT	MASS %	MW	LBMOL*	MOLE %
Chlorobenzene	58	112.56	0.515	34.89
DDT	26	354.49	0.073	4.97
Water	16	18.02	0.888	60.14

*Basis: 100 lb of waste.

From Table (5-1), the average molecular weight of the waste mixture is

$$\overline{MW} = (0.3489)(112.56)+(0.0497)(354.49)+(0.6014)(18.02)$$

$$= 67.73 \text{ lb/lbmol} \qquad (5\text{-}1)$$

The overall net heating value of the waste is calculated from the componential NHVs.

$$NHV = (0.3489)(714{,}361)+(0.0497)(1{,}580{,}000)+(0.6014)(-10519)$$

$$= 321{,}440 \text{ cal/gmol} = 578{,}600 \text{ Btu/lbmol} \qquad (5\text{-}2)$$

Using the average molecular weight of the waste mixture,

$$NHV = 578{,}600/67.73 = 8543 \text{ Btu/lb} = 4746 \text{ cal/g}$$

Next, the componential waste composition is converted to the elemental waste composition. This is done by summing the contributions of each of the four elements involved. For carbon, for example, 100 lb of waste produces 0.515 lbmol of chlorobenzene, 0.073 lbmol of DDT and 0.888 lbmol of water. The number of lbmoles (or lbatoms) of carbon in the waste is therefore

from C_6H_5Cl: (6 lbatom C/lbmol)(0.515 lbmol) = 3.1 lbatom C

from DDT: (14 lbatom C/lbmol)(0.073 lbmol) = 1.0 lbatom C

from Water: (0 lbatom C/lbmol)(0.888 lbmol) = 0.0 lbatom C

TOTAL = 4.1 lbatom C

The elemental waste composition is given in Table (5-2).

To determine the combustion products using stoichiometric air, Eqn. (3-2) may be used. The subscripts of the elements of the waste mixture are from Table (5-2) and are based on 100 lb of the waste.

$$C_{4.119} H_{5.013} O_{0.888} Cl_{0.882} + 4.71\ O_2 + 17.71\ N_2$$

$$\text{-----}> 4.119\ CO_2 + 2.07\ H_2O + 0.882\ HCl + 17.71\ N_2 \quad (5\text{-}3)$$

With 75% excess air, Eqn. (5-3) becomes

$$C_{4.119} H_{5.013} O_{0.888} Cl_{0.882} + 8.24\ O_2 + 31.00\ N_2$$

$$\text{-----}> 4.119\ CO_2 + 2.07\ H_2O + 0.882\ HCl + 3.53\ O_2 + 31.00\ N_2 \quad (5\text{-}4)$$

The flue gas composition based on Eqn. (5-4) is given in Table (5-3).

Table (5-2). Waste elemental composition.

ELEMENT	MASS %	LB/HR	AW	LB-ATOM*	ATOM %
CARBON	49.47	3181	12.01	4.119	37.78
HYDROGEN	5.05	325	1.01	5.013	45.98
OXYGEN	14.21	914	16.00	0.888	8.15
CHLORINE	31.27	2011	35.45	0.882	8.09
TOTAL	100.00	6430		10.901	100.00

*Basis: 100 lb waste.

Table (5-3). Initial flue gas composition.

COMPONENT	LBMOL*	MOLE %	LB*	MASS %
CARBON DIOXIDE	4.12	9.90	181.33	14.72
WATER (g)	2.07	4.98	37.27	3.03
NITROGEN	31.00	74.52	868.32	70.48
OXYGEN	3.53	8.49	112.98	9.17
HYDROGEN CHLORIDE	0.88	2.11	32.17	2.60
TOTAL	41.60	100.00	1232.07	100.00

* Basis: 100 lb waste mixture

The average value of the temperature in the incinerator is estimated using Eqn. (2-43). Since the waste mixture contains a significant amount of water, the suggested value of 0.30 for the heat capacity is replaced by 0.35. The value of EA in this problem is 0.75.

$$T = 60 + (NHV/0.35)/[1+(1+EA)(7.5\times10^{-4})(NHV)]$$

$$= 60 + (8543/0.35)/[1+(1+0.75)(7.5\times10^{-4})(8543)]$$

$$= 2059\ ^{o}F = 1399\ K \qquad (5\text{-}5)$$

The chemical reaction equilibrium constant for the HCl/Cl_2 distribution,

$$2\ HCl + 0.5\ O_2 = H_2O + Cl_2$$

is obtained from Eqn. (2-59).

$$\ln K = 7048.7/T + 0.0151 \ln T - 9.06x10^{-5}\ T$$
$$- 2.714x10^{4}\ T^{-2} - 8.09$$
$$= (7048.7/1399) + 0.0151(\ln 1399) - 9.06x10^{-5}(1399)$$
$$- 2.714x10^{4}/(1399)^{2} - 8.09$$
$$= - 3.083$$

$$K = 0.0458 \qquad (5\text{-}6)$$

From Eqn. (2-60),

$$K = P(Cl_2)\ P(H_2O)\ /\ P(HCl)^{2}\ P(O_2)^{1/2}$$

The partial pressures of the components before any Cl_2 is formed can be obtained directly from Table (5-3). Since the incinerator is assumed to operate at 1 atm pressure, the partial pressure for each component (in atm) is equal in magnitude to its mole fraction. Therefore, for carbon dioxide, for example,

$$P(CO_2) = [N(CO_2)/N_{TOT}]\ P_{TOT} = [0.0990\ mol/1.000\ mol](1.0\ atm)$$
$$= 0.0990\ atm \qquad (5\text{-}7)$$

The partial pressures for each flue gas component before and after the Cl_2 formation are represented in Table (5-4). The term z represents the number of moles of Cl_2 formed starting with one mole of flue gas before equilibrium.

Substitution of the partial pressures and the equilibrium constant into Eqn. (2-60) yields

$$0.0458 = (z)(0.0498+z)/(0.0211-2z)^{2}(0.0849-0.5z)^{1/2}\ N^{1/2} \qquad (5\text{-}8)$$

where N = 1.0000 - 0.5z

Note that the term N, which appears in each partial pressure, has been factored out and appears only once in Eqn. (5-8).

Table (5-4). Flue gas partial pressures.

Component	P(i) before Cl_2 formation	No. of moles of component*	P(i) at equilibrium*
CO_2	0.0990	0.0990	(0.0990)/N
H_2O	0.0498	0.0498+z	(0.0498+z)/N
N_2	0.7452	0.7452	(0.7452)/N
O_2	0.0849	0.0849-0.5z	(0.0849-0.5z)/N
HCl	0.0211	0.0211-2z	(0.0211-2z)/N
Cl_2	0.0000	z	(z)/N
TOTAL	1.0000	1.0000-0.5z	1.0000

* Basis: 1 mole of flue gas before Cl_2 formation

** N = total number of moles of flue gas at equilibrium
= 1.0000 - 0.5z

Solving Eqn. (5-8) by trial and error,

$z = 1.164 \times 10^{-4}$ atm

The corrected partial pressures and compositions are given in Tables (5-5) and (5-6).

Table (5-5). Equilibrium flue gas partial pressures.

Component	Partial pressure (atm)
CO_2	0.0990
H_2O	0.0499
N_2	0.7452
O_2	0.0848
HCl	0.0209
Cl_2	1.164×10^{-4}
TOTAL	1.0000

Table (5-6). Equilibrium flue gas composition.

COMPONENT	LB*	MASS %	LBMOL*	MOLE %
CARBON DIOXIDE	181.33	14.72	4.12	9.90
WATER (g)	37.42	3.04	2.08	4.99
NITROGEN	868.32	70.48	31.00	74.52
OXYGEN	112.94	9.17	3.53	8.48
HYDROGEN CHLORIDE	31.64	2.57	0.87	2.09
CHLORINE	0.34	0.03	0.005	0.01
TOTAL	1231.99	100.00	41.60	100.00

*Basis: 100 lb waste mixture

A rigorous calculation for the temperature is now performed by means of an enthalpy balance. The required heat capacity data are provided in Table (5-7). These data were taken from Table (2-4); the second column is from Table (5-6).

Table (5-7). Heat capacity vs. temperature data.*

Compound	gmole per gram of waste	a	$bx10^3$	$cx10^{-5}$
CO_2	4.12×10^{-2}	10.57	2.10	-2.06
H_2O	2.08×10^{-2}	7.30	2.46	+0.00
N_2	31.00×10^{-2}	6.83	0.90	-0.12
O_2	3.53×10^{-2}	7.16	1.00	-0.40
HCl	0.87×10^{-2}	6.27	1.24	+0.30
Cl_2	0.05×10^{-3}	8.85	0.16	-1.80

* $C_p = a + bT + cT^{-2}$ (cal/gmol-K)

From Table (5-7), for one gmole of waste,

$$\Delta a = 3.01$$

$$\Delta b = 4.63 \times 10^{-4}$$

$$\Delta c = -1.337 \times 10^{4}$$

When the waste and air are near room temperature and the combustion can be assumed to be adiabatic, Eqn. (2-42) reduces to

$$0 = NHV + \Delta H_{FG}$$

where $\Delta H_{FG} = \Delta H_{E2} + \Delta H_P$

The NHV for the waste mixture was evaluated earlier in Eqn. (5-2) as -4746 cal/g. The negative sign on the NHV is required when using Eqn. (2-42). The convention used in this equation is that heat lost by the system (waste, air and products) is negative and heat gained is positive. ΔH_{FG} is evaluated using Eqn. (2-12) with T_0 = 298 and T_1 = T (incinerator temperature).

$$0 = (-4746) + 3.01(T-298) + (0.5)(4.63x10^{-4})(T^2-298^2)$$
$$- (-1.337x10^4)[(1/T)-(1/298)] \qquad (5-9)$$

Solution of Eqn. (5-9) by trial and error yields

$$T = 1677\ K = 2559\ ^oF$$

Based on this revised temperature, the equilibrium constant for the HCl/Cl_2 equilibrium, calculated from Eqn. (2-59), becomes

$$K = 0.0195\ atm^{-1/2}$$

The partial pressures of the flue gas components are recalculated as before using the new value of K. The final flue gas composition is shown in Table (5-8). Because of the small amount of chlorine formed, the results in this table do not differ dramatically from those shown in Table (5-6).

Table (5-8). Final equilibrium flue gas composition.

COMPONENT	LB*	MASS %	LBMOL*	MOLE %
CARBON DIOXIDE	181.33	14.72	4.120	9.90
WATER (g)	37.27	3.03	2.068	4.97
NITROGEN	868.32	70.48	31.001	74.52
OXYGEN	112.98	9.17	3.531	8.49
HHDROGEN CHLORIDE	32.02	2.60	0.878	2.11
CHLORINE	0.15	0.01	0.002	0.01
TOTAL	1232.07	100.00	41.600	100.00

*Basis: 100 lb waste-fuel mixture.

From the results in Table (5-8), the flue gas mass, molar and volumetric flow rates are calculated.

$$\dot{m} = (1232.07\ lb/100\ lb_w)(6430\ lb_w/hr) = 79{,}220\ lb/hr \quad (5\text{-}10)$$

$$\dot{n} = (41.600\ lbmol/100\ lb_w)(6430\ lb_w/hr) = 2{,}675\ lbmol/hr \quad (5\text{-}11)$$

$$\dot{V} = \dot{n}RT/P = (2675)(0.7302)(2559+460)/(1)$$
$$= 5.897\text{x}10^6\ ft^3/hr = 98{,}290\ acfm \quad (5\text{-}12)$$

The volume of the incinerator is calculated from Eqn. (3-8) using the combustion heat rate at 60°F (q) and a heat release rate (q_H) of 25,000 Btu/hr-ft^3.

$$V = q/q_H = \dot{m}\ (NHV)/q_H = (6430)(8543)/(25{,}000)$$
$$= 2{,}197\ ft^3 \quad (5\text{-}13)$$

Since L/D is specified as 4.0, the incinerator diameter and length may be determined by solving Eqns. (5-14) and (5-15) simultaneously.

$$V = 2197 = (\pi D^2/4)L \quad (5\text{-}14)$$

$$L/D = 4.0 \quad (5\text{-}15)$$

The results are D = 8.88 ft and L = 35.50 ft.

The overall residence time (t) and superficial velocity (v) are calculated from Eqns. (3-5) and (3-8).

$$t = V/\dot{V} = 2197/(98290/60) = 1.34\ sec \quad (5\text{-}16)$$

$$v = V/A = (98290/60)/[\pi(8.88)^2/4] = 26.5\ ft/sec \quad (5\text{-}17)$$

These values of t and v are reasonable.

Tables (5-9) and (5-10) present these and some other results calculated by the *HWI* program.

Table (5-9). Flue gas data.

Mass flow rate, lb/hr	79222.
Volumetric flow rate, acfm	98292.
Volumetric flow rate, scfm (60 °F)	16928.
Volumetric flow rate, dscfm (60 °F)	16086.
Molar flow rate, lbmol/hr	2675.
Temperature, °F	.2559.
Sulfur trioxide concentration, ppm	0.0
Chlorine concentration, ppm	50.9
HCl mass flow rate, lb/hr	2058.6
Particulate loading, gr/acf	0.00
Particulate loading, gr/dscf -- corrected to 50% excess air	0.00

Table (5-10). Preliminary design data.

Heat generation rate, Btu/hr	0.5493E+08
Volume, cu ft	2197.
Diameter, ft	8.9
Length, ft	35.5
L/D ratio	4.00
Superficial velocity, ft/sec	26.5
Residence time, sec	1.34

References

1. Theodore, L., Reynolds, J., *Introduction to Hazardous Waste Incineration*, Wiley-Interscience, New York, 1987 (drawn in part from).

2. *CRC Handbook of Chemistry and Physics*, 51st ed., Chemical Rubber Publishing Co., 1971.

3. Smith, J.M. and Van Ness, H.C., *Introduction to Chemical Engineering Thermodynamics*, 3rd ed., McGraw Hill, 1971.

4. Fogler, H.S., *The Elements of Chemical Kinetics and Reaction Calculations*, Prentice Hall, 1986.

5. Personal Notes: L. Theodore, J. Reynolds.